Mathematics Galore!

Mathematics Galore!

Masterclasses, Workshops, and Team Projects in Mathematics and its Applications

C. J. BUDD
Department of Mathematical Sciences, University of Bath

C. J. SANGWIN
School of Mathematics and Statistics, University of Brimingham

OXFORD
UNIVERSITY PRESS

OXFORD
UNIVERSITY PRESS

Great Clarendon Street, Oxford OX2 6DP

Oxford University Press is a department of the University of Oxford.
It furthers the University's objective of excellence in research, scholarship,
and education by publishing worldwide in

Oxford New York

Athens Auckland Bangkok Bogotá Buenos Aires Calcutta
Cape Town Chennai Dar es Salaam Delhi Florence Hong Kong Istanbul
Karachi Kuala Lumpur Madrid Melbourne Mexico City Mumbai
Nairobi Paris São Paulo Shanghai Singapore Taipei Tokyo Toronto Warsaw

with associated companies in Berlin Ibadan

Oxford is a registered trade mark of Oxford University Press
in the UK and in certain other countries

Published in the United States
by Oxford University Press Inc., New York

First published 2001

A catalogue record for this book is available from the British Library

Library of Congress Cataloging in Publication Data

Budd, C. J. (Christopher J.)
Mathematics galore!: Masterclasses, workshops, and team projects in mathematics and
its applications / C. J. Budd and C. J. Sangwin.
Includes bibliographical references.
1. Mathematics–Study and teaching (Secondary) I. Sangwin, C. J. (Christopher J.)
II. Title.

QA11 .B856 2001 510′.71′–dc21 00-066544

ISBN 0 19 850769 0 (Hbk)
0 19 850770 4 (Pbk)

Typeset by Newgen Imaging Systems (P) Ltd., Chennai, India
Printed in Great Britain
on acid-free paper by
TJ International Ltd,
Padstow, Cornwall

Contents

Introduction

Where we start

If you were to ask enthusiastic mathematicians whether or not they thought that mathematics was a golden thread in life's rich tapestry, then the chances are that they would say that it was the whole tapestry, together with most of the room in which it is displayed. Mathematics is amazing in its power and beauty and the way that it has applications in so many areas of human endeavour. Mathematics is also central to much of modern technology and its applications to this will become increasingly important in the new millennium. But mathematics is ever changing and growing, and it will only continue to develop and thrive if fresh minds and new ideas are applied to it. The need to stimulate young people to become interested in mathematics led to the creation of the Royal Institution Mathematics Masterclasses.

The aims of a mathematics masterclass and of this book are to

- enthuse
- educate
- inspire
- challenge

audiences of young people, their parents and teachers, with the wonder, excitement, power, beauty, and relevance of modern mathematical ideas. This is done through a mixture of presented material, problems (often tackled in teams), practical work, close interaction with professional mathematicians, ideas to take home, and field trips to interesting mathematical locations. Above all masterclasses are *fun* and our aim is that everyone attending should leave fizzing with the excitement of the mathematics that they have been doing. Hopefully, reading this book will give you a similar experience.

Since 1981, inspired by Sir Christopher Zeeman FRS, the Royal Institution of Great Britain has sponsored the creation of mathematics masterclass groups all over the UK. These groups present exciting mathematical classes to 13–14 year olds who are encouraged to come by their schools. The book *Mathematical Masterclasses, Stretching the Imagination* edited by Prof. M. Sewell and published by Oxford University Press, is a collection of classes given by the Reading and Berkshire group.

The Bristol and Bath Masterclass group was founded in 1990 and ever since has run regular Saturday morning workshops for students aged 13–14, all of whom come from local schools. The demand is such that three sets of eight classes are run each year, attracting around 250 young people in total annually. These classes have been a team effort from the Universities of Bath, Bristol, and the University of the West of England, in close and splendid partnership with many teachers. They have been supported for several years by a grant from Zurich Insurance. Details of the annual programme can be found on: `http://www.bath.ac.uk/RIBBM`

The Royal Institution of Great Britain coordinates masterclasses all over the UK. Information on these can be found on the Royal Institution Web site at `http://www.ri.ac.uk/`

If you do not have access to a masterclass group and/or would like to learn some new mathematics and try your hand at mathematical puzzles and games, then we strongly recommend the online mathematics club NRICH which is supported by the Royal Institution and the Mathematics for the Millennium Project based at the University of Cambridge. NRICH can be found at `http://nrich.maths.org.uk/`

The chapters in this book are all based on classes which have been presented to the Bath–Bristol group and they have been tested by some very demanding and lively audiences. The aim has *not* been to stick to material which is likely to be met in a typical school syllabus. Rather, we have let ourselves explore some areas of mathematics a long way from traditional courses. From mazes to folk dancing and from knitting to castle design. We hope that you will find them fun and challenging and they will lead you to look at the world with a new mathematical perspective.

How does a masterclass work?

Faced with an audience of 50–100 dynamic teenagers, often with very different mathematical backgrounds and abilities, how do you present exciting mathematics for two and a half hours on a Saturday morning? The answer is *get everyone involved*. A good masterclass is a mixture of taught material broken up by group workshop sessions in which everyone (young people, teachers, and their parents) works together through problems or practical projects. These workshops should have a variety of material to suit everyone, regardless of their ability or experience. A typical format of a masterclass which has been found to work well is

- 10.00 a.m.–10.30 a.m. Arrival, introduction and first talk session
- 10.30 a.m.–11.00 a.m. First workshop session
- 11.00 a.m.–11.30 a.m. Second talk session
- 11.30 a.m.–11.45 a.m. Refreshments
- 11.45 a.m.–12.15 p.m. Second workshop session
- 12.15 p.m.–12.30 p.m. Summing up session
- 12.30 p.m. Depart

The workshops are best arranged with the participants working in small groups with 'experts' and parents helping out. These experts can be university faculty, teachers, and students studying mathematics at university who are also keen to work with young people. A good masterclass should have more than enough material for a workshop, leading to follow-up later both at home and at school. No masterclass will work without careful planning beforehand, especially if it is for a large number, and close cooperation is needed between the presenter and those assisting with the workshop sessions.

What is this book all about?

The aim of this book is to convey some of the fun, excitement, and understanding of mathematics that is part of the masterclass experience. Each of the chapters gives a self-contained collection of material suitable for a masterclass session and the chapters cover a wide variety of topics. There is explanatory text, material for the workshops themselves, and also additional material which allows the ideas in the masterclass to be extended further.

This book is meant both to be read and to be used! The masterclass could either be part of a series running over several weeks or instead could be a one-off if you want some ideas to do something a bit different in a class.

However, it is not intended that the book should only be used in this way. We hope that everyone of all ages, including young people, parents, teachers, or anyone interested in mathematics, will find it fun and exciting to read the different chapters and the exercises these contain, and to discover more about mathematics by doing so.

The main body of the text of each chapter is core material suitable for a general reader from the age of about 13 upwards. This is material that can either be presented in a class or studied on its own. (We should say from experience that some material, when suitably presented, also works for primary school audiences from the age of 5 upwards.) The aim of the core part of each chapter is to develop key mathematical ideas and to place them in the context of novel, interesting, and unexpected applications to real-world problems. For example, in the chapter on mazes you learn how to amaze your friends by cracking a maze and also learn about the theory of networks. In the chapter on sundials, as well as constructing your own sundial, you also learn some trigonometry.

Many new mathematical ideas are presented in the core material, and the key process of taking a problem (often coming from a real-world application), solving it, abstracting the mathematical idea, and then solving many other apparently quite different problems, has been a driving theme in the material and topics that we have selected. We hope to give you enough mathematics in this section to motivate and inspire you, without swamping you with detail.

At the end of the core material are a set of problems listed as First Session and Second Session. These are divided up in correspondence to the two workshops described in the timetable above. These problems are best solved in small groups but you can also tackle them on your own. We come to mathematics from many different angles and backgrounds and the problems reflect this. Some are very open ended and require creative and imaginative thinking. Others are more challenging mathematically and need some technical thinking to master. There are many more problems than can be tackled in a single session, so take your pick and do the ones that you find interesting. Please copy these sheets and use them.

In history you go on field trips, in geography you go on field trips. Why not go on mathematical field trips? We have tried them – and they work. After the problems we give some ideas for good trips that you can take a whole class to or can form the basis of a family outing. (During the course of the research for this book the Budd family has visited a large number of castles!) Looking at the real world through mathematical eyes can be a lot of fun. Even the coach journey can be used to advantage, by providing everyone with problem sheets to puzzle over as they travel.

After the problems used in conjunction with the two masterclass sessions, we have provided some further problems. These are much longer and more open ended. They are meant to really stretch you. This section is designed for teachers, parents and/or advanced school mathematicians. It is more detailed than the main body of the text and is intended to give you something to really get your teeth in to and to develop the masterclass material. We also provide answers to the problems and suggest some further reading and references.

We encourage you to use your imagination as you read this book and to be creative both in learning and presenting mathematics.

The classes

Chapter 1 Amazing mazes
This chapter links network theory to the study of mazes. We give a short history of mazes, starting with the Minotaur. We prove that there are some mazes you can crack by keeping your hand on a hedge. We introduce some ideas of network topology and consider the bridges of Königsberg. This leads to a general method for solving all mazes. Workshop material is then given, together with details of mazes which can form the basis of a field trip. We finish with more advanced material about the theory of networks.

Chapter 2 Dancing with mathematics
This chapter links group theory to the study of patterns in dancing, bell-ringing, and knitting. The first part looks at the symmetries of the square and derives some basic properties of the dihedral group. These properties are connected in a natural way to patterns in dancing and they are used to devise some new dances. The methods are then extended to look at related patterns in bell-ringing and knitting. The workshop material applies these methods and also encourages the reader to try some dancing. The more advanced material is a discussion of group theory in more detail with reference to some specific groups.

Chapter 3 Sundials: how to tell the time without a digital watch
This chapter motivates some results in trigonometry by linking them to the problem of telling the time. It starts with a history of different ways of telling the time and goes on to look at the way that the sun appears to move around the sky, giving formulae for the angle that it makes with the horizon at different times of the year. This theory is then used to design equatorial, horizontal, vertical, and analemmatic sundials. The time shown on a sundial is compared with the time shown on a watch and the equation of time is explained. The exercises include some practical projects for constructing sundials. In the further problems some more advanced trigonometry is introduced, to derive the formula for the angle of the sun and also to find out the formula for the dial of the horizontal sundial.

Chapter 4 Magical mathematics
We demonstrate in this chapter that although mathematics is a logical subject, it still contains many surprises. We start by looking at some remarkable mathematical theorems (mostly involving π) that are in every sense magical and surprising. We then show how different theorems can lead to mysterious magic tricks involving watches, cards, and mind reading. First the magic trick is introduced and then the mathematics behind it is explained. You are encouraged to try these out in a magic show. In the further problems we look at surprises arising in probability where our common sense can easily lead us astray. In particular we look at a lottery based on choosing 6 balls out of 49. We consider the probabilities of various events and compare actual probabilities with people's perceptions of probabilities as evidenced by their choices

of numbers. As numbers are often chosen in relation to people's birthdays we look at various surprises connected with birthdays.

Chapter 5 Castles: mathematics in attack and defence
This chapter shows how geometry and symmetry play a vital role in the design of castles. We start by looking at ancient hill forts which are often circular in form. A reason for this is that the greatest area for a given perimeter is enclosed by a circle. A proof of this theorem is given which involves simple concepts and is well suited to a team effort (young people in a circle making up the perimeter). We also look at the design of entrances to hill forts and link these to the design of mazes in Chapter 1. We then look at later designs of castles. Questions are asked, such as where is a good place to put arrow holes and why do castles have turrets? A detailed study is made of the geometry of Deal castle in Kent. Finally we look at more modern fortifications based upon the pentagon with bastions, and see why this is such a good design. In the exercises you are encouraged to design a castle based upon the principles mentioned earlier and to consider the importance of symmetry. In the further problems we look at properties of some of the regular shapes encountered in the design of castles and also give a more detailed discussion of the relation between the length of a curve and the area it encloses – which links to fractals and the entrances of castles.

Chapter 6 How to be a spy: the mathematics of codes and ciphers
This chapter relates modular arithmetic, statistics, and number theory to code breaking. It starts with a historical account of code breaking, concentrating on the breaking of the Enigma cipher. It then introduces the Caesar cipher and shows how this is related to modular arithmetic. Various results are established about modular arithmetic and they are used to help crack the cipher. More general simple substitution ciphers are introduced and it is shown how to crack these with the use of statistics. We look at multiple alphabet ciphers and show how they can be solved using a Viginère square – linking in with coordinate geometry. Finally we look at transposition ciphers and relate these to prime factorizations of numbers. In the workshop sessions we study further properties of modular arithmetic and give lots of examples of codes and ciphers to crack. In the further problems we look at the modern RSA cipher which is based upon properties of the prime numbers.

Chapter 7 What's in a name?
A dictionary defines a number to be: (i) a quantity or amount, and (ii) a symbol representing this. These two ideas are entirely different and this chapter will examine this and other interesting things about numbers themselves. This includes a historical discussion of systems for writing numbers: Babylonian cuneiform, Egyptian hieroglyphics, Roman numerals, and Mayan scripts, etc.; the history of number systems together with their advantages and disadvantages; the nature of zero (and nought); number bases and negative number bases. In the workshops we ask the students to write and interpret sums using different number systems and also to practise some calculations using number bases and negative number bases.

Chapter 8 Doing the sums
Practical calculation is just as important as theory, and being able to accurately and efficiently perform arithmetic has been vital to the development of science. First we look at logarithms and rapidly progress to the slide rule. Now neglected in favour of a calculator, we firmly believe that a basic understanding of these concepts is very helpful. The workshop sessions reflect the practical theme of this chapter and students practise using the techniques they have learned. In the further problems we expand on some of the ideas and hint at results from higher mathematics.

Acknowledgments

We would particularly like to thank:

The University of Bath Department of Mathematical Sciences, the Members of the Bristol and Bath Masterclass Committee, the Royal Institution of Great Britain and the Mathematics for the Millennium team; Aaron Wilson, for help and advice with typesetting; Martin Brown, Alistair King, Ann Sangwin, Jill Budd, Vanessa FitzGerald for carefully reading chapters of the book; Dr Eleanor Robson, Oriental Institute Oxford, for sending us on a mathematical treasure hunt of the Ashmolean museum; Lindsay Heyes for many interesting discussions on the history and theory of mazes; David Brown for much help and advice with sundialing and also providing a photo of his sundial in Figure 3.9; Dr Ossama Alsaadawi for providing the translation of the Egyptian text in Figure 7.5.

Figure 1.6 is reproduced with kind permission of The Dean and Chapter of Hereford Cathedral and the artist Dominic Harbour. Figure 1.13 is taken from a poster that was part of the 'Posters in the London Underground' project, designed by Dr. Andrew Burbanks at the Isaac Newton Institute for Mathematical Sciences, and reproduced here with kind permission. Figure 1.18 is reproduced with kind permission of Lindsay and Edward Heyes. Also we thank Christina Smith and the pupils of St Mary's School Calne for permission to reproduce a picture of their sundial as Figure 3.14. We would like to thank Bryony and Jeremy Budd and William Hughes who appear in Figures 5.3, 5.22, and 5.26 and Sue Budd, who took the photographs. Figure 5.9 is reproduced with kind permission of Dorset County Council. Figure 6.1 is reproduced with permission of the Science Museum and Society Picture Library. Figure 7.5 is reproduced with permission of Lotos Film.

Chapter 2 is based on an earlier article by the authors for the University of Bristol magazine *Nonesuch*, and is used here with kind permission.

We would like to thank the many other people who have helped with useful encouragement and especially all the participants of masterclasses on whom we've tried out this material.

Lastly and most importantly we would both like to thank our families for their support and assistance in so many ways during the work on this project.

1 Amazing mazes

1.1 Introduction

Unlike most other mathematical problems, the study of mazes and labyrinths takes us into the dark territory of murder, suicide, adultery, passion, intrigue, religion, and conquest. Readers of a nervous disposition should proceed rapidly to the next chapter. However we encourage you to stay on, as this study will take us from the dawning of mathematical thought to the modern technology of digital electronics with many interesting diversions along the way. Solving mazes has traditionally been a source of recreation for all ages, but the mathematical techniques developed to accomplish the solution have applications to many important problems.

1.2 The story of the Minotaur

Mazes are very ancient and appear many times in history. The first hero of our story is the inventor Daedalus, to whom, amongst several discoveries, is given the credit for inventing sailing ships and manned flight. According to ancient legend, Daedalus constructed the so-called 'Cretan Labyrinth' in Knossos, the capital of the Island of Crete. This was supposed to be an early and highly complex maze especially built to house the legendary Minotaur. The Minotaur was a fearsome creature, half-man and half-bull, and the product of an uncertain liaison between a bull and Pasiphae, the queen of King Minos of Crete. We can see him in Figure 1.1. Now, Androgeus, the son of Minos, had been killed by the Athenians and, in revenge, Minos forced the Athenians to send seven young men and seven young women every nine years to Crete. They left Athens amidst great sorrow, in a boat with black sails. On arrival at Crete they were taken one by one into the Labyrinth in which (of course) they became quickly and hopelessly lost. Thus they became easy prey for the Minotaur and after a short interval were killed and eaten. Naturally this state of affairs did not appeal greatly to the Athenians and in an attempt to remedy the situation Theseus, son of Aegeus, King of Athens, resolved to sail with the boat for Crete in place of one of the seven young men. Before sailing, Theseus made a deal with his father so that if he was successful they would sail back with the usual black sails of the ship changed to white.

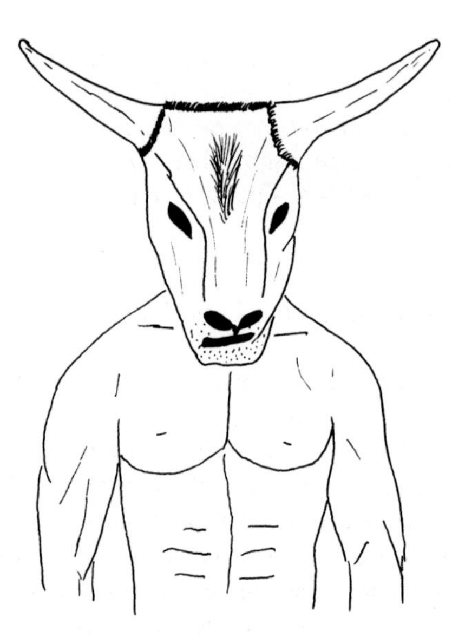

Figure 1.1: The Minotaur.

On arrival at Crete the ship landed in a bay close to Minos' palace, and, as luck would have it, King Minos' daughter Ariadne was tending sheep by the bay. On meeting Theseus, in the tradition of all good stories, she fell deeply in love. Ariadne vowed she would aid Theseus in his quest against the Minotaur, providing him both with a sword and, more importantly, a clue to solving the Labyrinth. Her solution was to provide Theseus with a thread which Theseus would unravel as he entered the Labyrinth, and to get out all he had to do was to rewind the thread bit by bit. It is interesting to note that Ariadne gave Theseus a way to get *out* of the Labyrinth but not a way to get into the centre and to find the Minotaur. This is a much harder problem and was not really solved until the eighteenth century (by Euler), long after Crete had been destroyed. Notwithstanding the fact that he only knew the way out, Theseus entered the Labyrinth successfully and in an epic struggle (blood, etc.) slew the Minotaur. Using Ariadne's thread he then re-traced his steps, rescued the Athenians and, taking Ariadne with him, escaped in the ship now bound for Athens. So far so good, but now Theseus makes two slight errors of judgement. The ship was in no great hurry to get back to Athens and called on several islands on the way back. Landing at the Island of Naxos, the Athenians performed various dances showing Theseus' path through the Labyrinth. They were feasted and entertained by the local citizens and sailed off in good spirits, only realizing after they had departed that they had left Ariadne (who was now pregnant) behind. Ariadne died, either of a broken heart or in childbirth and never saw her homeland again. Furthermore, on his return to Athens, Theseus sadly forgot to change the sails on his ship from black to white. King Aegeus who had been desparately scanning the horizon awaiting Theseus' return finally caught site of a ship which bore black sails. He could only conclude that Theseus had been vanquished. In his despair he threw himself into the (now called) Aegean sea. Thus Theseus had caused by his gross negligance, the death of both his lover and his father. Fortunately for his reputation this behaviour is just par for the course if you are a Greek hero!

1.3 The mathematics of the Cretan Labyrinth

Early in the twentieth century, the British archaeologist Sir Arthur Evans excavated the site of King Minos' palace in Crete and found much archaeological evidence to support the legend of the Minotaur. In particular the walls were covered in images of young men and women in athletic games involving bulls – the Minoans evidently had a bull culture. Furthermore the palace had many double axes displayed in the halls – the Greek name for such an axe is *labrys* and it is possible that the word 'labyrinth' comes from this. However there was no direct evidence for a maze in the palace although the walls are covered in complicated maze-like patterns and the rooms of the palace are certainly in a very complex arrangement which would seem very much like a maze if you did not know your way around.

Although we don't have direct evidence in the form of buried walls for the shape of the Cretan Labyrinth, there is a traditional idea about its shape, and a very nice geometrical construction for drawing one. This gives us our first link between mathematics and mazes. You can draw this on paper, or if you are on a beach it looks very good drawn into the sand with the help of a stick. To draw a traditional Cretan Labyrinth, start with the cross and dots shown in Figure 1.2 (a).

Next add four arcs and label the points as shown in Figure 1.2 (b). Notice that the numbers alternate left and right round the square. Start by connecting 1 to 2. This is called the *seed* and is shown in Figure 1.3. Different seeds will give different labyrinths and you will have a chance to try different seeds in the exercises.

To continue, connect 3 to 4 and then 5 to 6. Can you spot the pattern of how to connect the lines up yet?

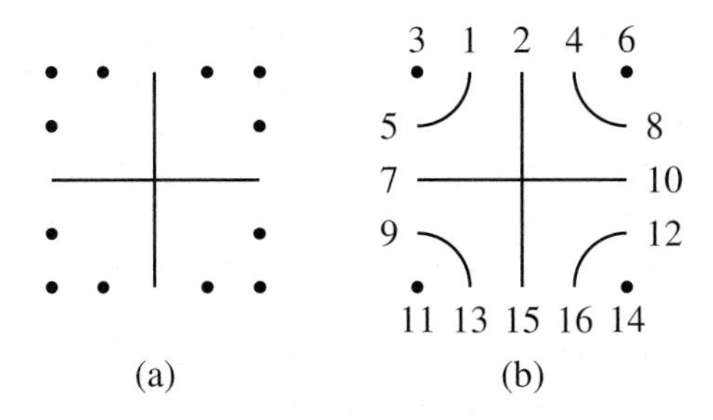

Figure 1.2: Starting the Cretan Labyrinth.

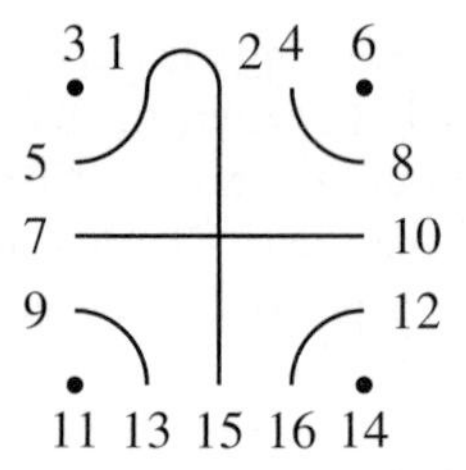

Figure 1.3: The seed of the Cretan Labyrinth.

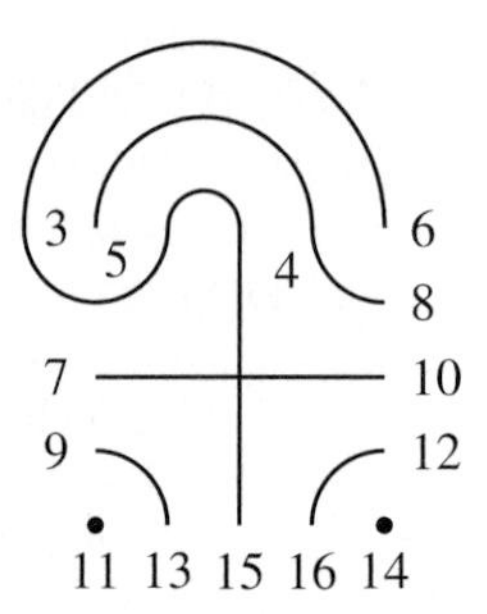

Figure 1.4: The next stage in the Cretan Labyrinth construction.

Look for the next two numbers and then join them up. The next two numbers are 7 and 8 and joining them up gives:

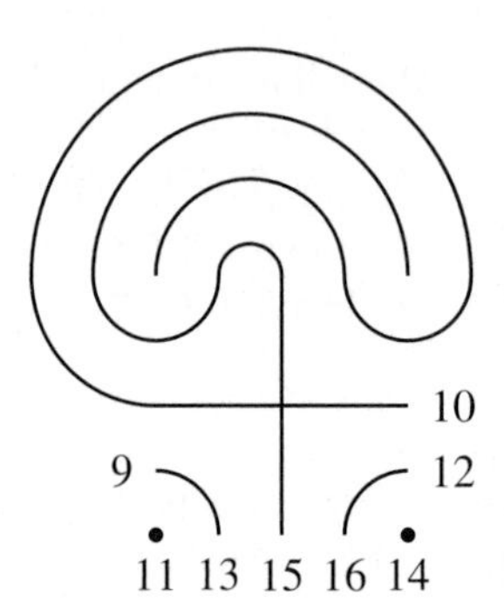

Exercise. *Now complete this picture. This gives you the* Cretan Labyrinth *shown in Figure 1.5.*

Exercise. *Try following the route from the entrance to the centre.*

This path is surprisingly long and in a full-size labyrinth it would have taken some time to get to the centre. In the exercises you can have a go at seeing how long it is. However, there is a further surprise in store. Although the path to the centre is

Figure 1.5: The Cretan Labyrinth.

very long there is only one way in and one way out! Theseus had to make no difficult decisions at all on his way to kill the Minotaur. Indeed, it was easy with this design to get to the centre and just as easy to get out again. In short there was no need for threads, Ariadne, broken hearts, suicide, or any of the other features of the story.

Exactly the same geometric pattern as the Cretan Labyrinth appears in many different cultures and it is quite a common artistic image. This is a good example of the universality of mathematical thought. (Another good example is the independent use of the pyramid for large buildings in both Egypt and Mexico.) Examples of similar designs have been found scratched into caves in Cornwall (possibly by visiting Phonecian seafarers or by visiting mathematicians in a moment of boredom), on Roman coins, and in pictures drawn by native American Indians. The pattern is of interest to mathematicians because it packs a very long path into a small space. (This is an early example of a mathematical object called a *space-filling curve*.) Later on in this book we will see how the idea of having a very long path in a small area is also useful in a military application when building the walls of castles. A similar pattern to the Cretan Labyrinth was supposed to have been used to build the walls of Jericho, and you can see the form of this in Chapter 5 when we look at castles. If you visit Hereford Cathedral in England you can see an early medieval map called the *Mappa Mundi*. On this the island of Crete is marked clearly, together with a labyrinth. This is shown in Figure 1.6. Interestingly this is a different design from the pattern we have described and was drawn by a monk using a pair of compasses. The hole made by the compass point is still visible on the map.

Using other seeds

Using different seeds when you start drawing leads to different labyrinth designs. An important feature of the Cretan Labyrinth is that there is only one entrance and only one route toward the centre. We can ask the (mathematical) question of whether all seeds lead to labyrinths with one entrance and route.

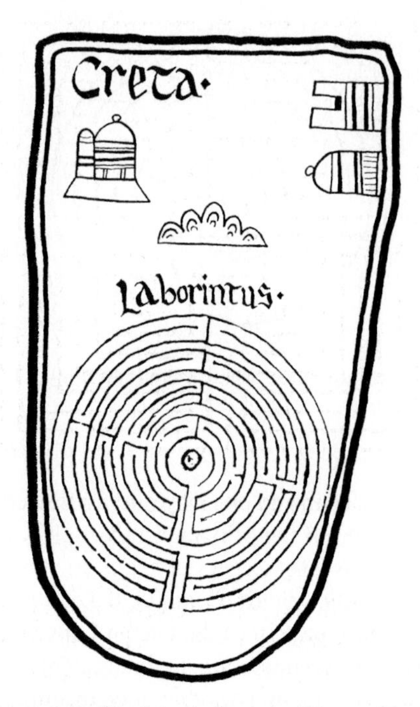

Figure 1.6: The labyrinth on the *Mappa Mundi*.

Although the Cretan labyrinth is the most common, and probably the oldest of these designs, there are lots of others that we can draw. To do this you need to start with a different seed. Let's try a different seed such as this one:

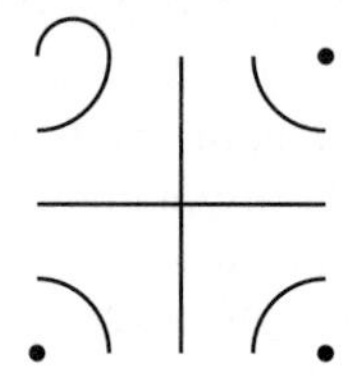

To draw the complete picture remember to alternate drawing the lines from left to right. The resulting labyrinth obtained from this seed looks like this:

Figure 1.7: A maze from a mosaic design at Calvatone, Italy.

There are some other seeds in the exercises for you to try out.

The Romans developed the Cretan Labyrinth pattern in the design of their mosaics, many of which can still be seen today. The Roman idea was to distort the labyrinth into a pattern with straight sides and to join four labyrinths together to give the mosaic. An example of a Roman mosaic is given in Figure 1.7.

Exercise. *See if you can work out how this labyrinth is made up of three copies of the same labyrinth. How does the fourth part differ?*

In the exercises you can have a go at making similar mosaic patterns.

1.4 The rise of the maze

The term 'labyrinth' is now generally associated with a construction that leads you from a starting point to a goal by taking you on a tortuous path, but requires no actual decisions. Your whole path is predetermined by the constructor of the labyrinth. Sacred sites were sometimes constructed as labyrinths – often out of turf – such as that on Rippon Common, Yorkshire, England, shown in Figure 1.8. This construction appealed to philosophers, who believed in the action of fate giving you an ultimate destiny which was entirely beyond control. However, following the Christianization of the Roman Empire, and the belief in the action of free will, a different form of construction came into being. This was the maze. In a maze intrepid travellers had to make a series of decisions, and their ultimate fates (in particular whether they reached the centre) relied upon the results of those decisions. Mazes were often built into the floors of churches and you were supposed to pray as you found your way towards the centre. The idea of the puzzle maze was developed during the Middle Ages and later

Figure 1.8: The plan of a turf maze on Rippon Common, Yorkshire, England.

into the celebrated hedge maze, often found in the grounds of stately homes. Many good stories are associated with these types of maze. Around AD 1176, King Henry II of England, who was married to Queen Eleanor, fell in love with (fair) Rosamund. In order to carry out his meetings with Rosamund in secret, Henry built a maze (called Rosamund's Bower) and as only he and she knew the way to the centre they were able to meet there in relative security. For a while this approach was successful, but sadly for Rosamund, Eleanor learnt the secret of the maze, entered it, and, finding Rosamund there, forced her to take a fatal dose of poison. (Sorry if this story seems rather familiar by now!)

The modern use of the hedge maze is now purely recreational. The puzzle is usually to find your way to the centre (and out again) starting from the entrance. A variation on this is for two of you to enter the maze at different times and either for you to try to find each other, or for one to hide in the maze and the other to find them. Many mazes around the world are open to the public and make a great day out. Examples include the Jubilee Maze Centre at Symonds Yat (which has a fine museum of mazes at its centre), and the maze at Longleat, which has bridges and changes its pathways during the day. If you get lost in this maze then clues are available. Perhaps the most famous public maze in England is the hedge maze at Hampton Court near London, which was constructed in AD 1690 and is still open to the public. In the early years of the twentieth century, the three men in the book *Three Men in a Boat* by Jerome K. Jerome, visited this maze. Despite having a sure-fire way of solving the maze the three men, under the leadership of a certain Harris, got splendidly lost. They even had a map but as Harris said 'there's not much point in having a map, if you don't know where you are in the maze'. During their time in the maze they gathered together a large crowd of similarly lost members of the public, including a woman with a baby and an inexperienced maze keeper. Only after some time, and the intervention of a more experienced maze keeper, were they eventually rescued. The full account of this story is very well worth reading. If you get the chance, visit the maze at Hampton Court and indeed Hampton Court itself. The maze plan is given in Figure 1.9.

Figure 1.9: The plan of the Hampton Court maze, with X marking the centre.

A list of some famous mazes around the world can be found in Table 1.1 at the end of this chapter. More examples of mazes can be found in the references and the related Web sites. Plans of some of these mazes are given in the exercises.

Mazes are found everywhere in modern life and have many important applications. An example of a modern maze is the road traffic network. This is an especially complex maze as it includes roads which only go one way and other roads which may be closed at certain times of the day. Anyone with a car (or a bicycle) has to solve the problem of finding their way through this maze as quickly as possible. You can now get computer programs to do this for you – and the ideas behind these problems came from attempts to find ways of solving mazes. Mazes are important to computers themselves. The links between computers on the World Wide Web form a very large maze linking all the users together. To find information quickly on the Internet requires the computer to solve a maze.

In fact computers would not work at all if it were not for a maze, for deep inside every computer chip is an arrangement of components and wires that is similar, mathematically speaking, in every way to the arrangements of hedges in a maze.

In the exercises we challenge you to think of other objects and constructions in the world that resemble mazes.

We hope that you are convinced that mazes are both fun and important. Now let's think how, as mathematicians, we could try to solve the puzzle of how to get to the centre of a maze (and out again) quickly and reliably.

1.5 How to solve a maze with your hand on a hedge

A maze has an entrance (or several) and a special point which we will call the centre. In a recreational puzzle this often has some special feature such as a statue or a fountain. However, mathematically the centre could be any point in the maze that we are trying to get to. To solve a maze you need to find any route from the entrance (or entrances) to the centre and back out again.

It is perhaps obvious to say that some mazes are easier to solve than others but there is a special class of mazes which can be cracked in a very easy way. Hampton Court is an example of such a maze and Harris did indeed have a method of cracking it. Here is Harris' method as quoted by him

> "solving a maze is easy, all you need to do is to *keep turning to the right*."

Suppose that we take the map of the Hampton Court maze given in Figure 1.9 and, as suggested, after entering the maze, keep turning to the right (in practice this means keeping your right hand always in contact with a hedge). Remarkably we find that after a few twists and turns the centre is reached. Furthermore, the same approach gets you out again. As Harris himself got lost in the maze, we can only conclude that he must have been too distracted by his companions to have followed his own advice.

We can be more precise and state Harris' method of solving the maze in terms of a mathematical theorem.

> Let M be any maze which we enter and keep turning to the right (perhaps by keeping your right hand on the hedge of the maze). Then either:
> - You will get to the centre, and on leaving the centre you will get out again; or
> - you will get out again, without reaching the centre.

The most important consequence of this theorem is that *under no circumstances* will you get lost or wander round aimlessly. In fact the theorem puts you into a win–win situation. Follow its advice and either you get to the centre – and win – or you get out again and can pretend (to your friends) that this was what you really meant to do all along.

> WARNING: Don't try this too often if you want to keep your friends.

We can show why this theorem is true by starting with a very simple maze and then making it more and more complicated by adding more and more hedges. We start with the maze below, which has an entrance A, a centre B, and not much else. It has a single hedge which separates the inside of the maze from the outside.

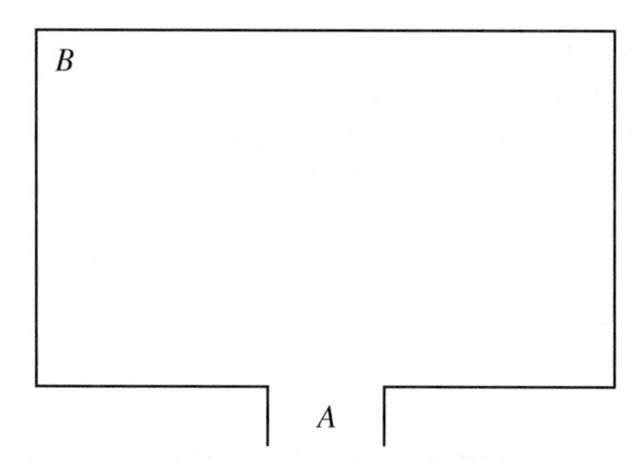

This is rather an easy maze – we would not get many crowds interested in solving this one, and the profits from our maze centre would be rather low. To reach the centre from the start we can obviously keep our right hand on the hedge.

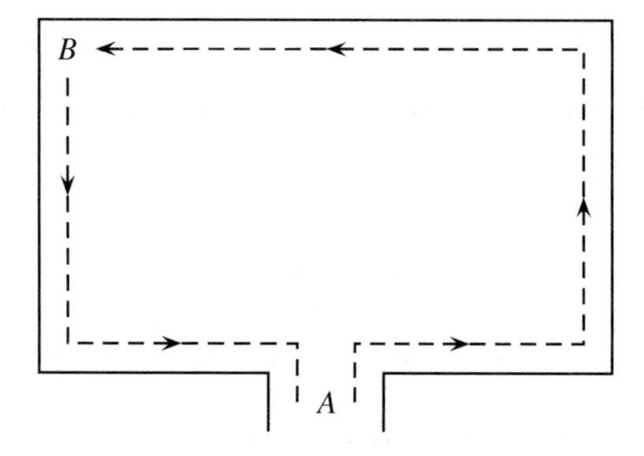

Once we get to the centre then it is equally obvious that keeping our hand on the hedge will get us out again. See the dotted path which shows you the route. Now suppose that we build a new hedge which is *attached to the outside hedge* at the point C on one side and the point D on the other side. This is indicated in the diagram below.

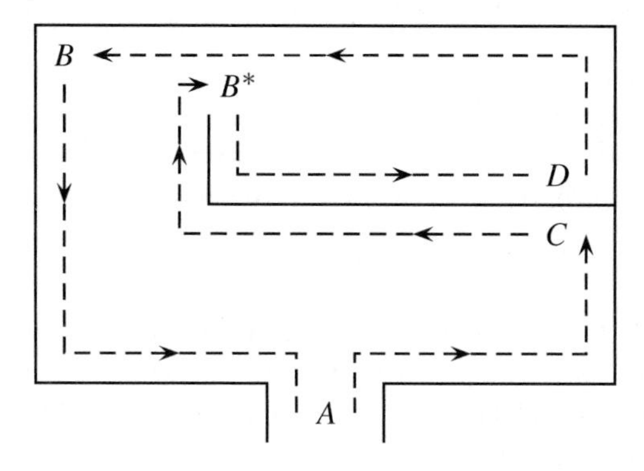

Starting from the entrance, follow the dotted line as before. Eventually you will get to the point C at the bottom of the hedge. If you now follow the hedge round with your hand you will go up one side, turn around the edge, and then will get to the point D. Now, to quote one of our parents *I know where I am now – I've been lost here before!* The point D is on our original dotted path. Now we can follow this path using our original method till we reach the centre. By doing this we have employed a very useful technique in mathematics:

> Transform a hard problem into a simpler one which we can solve more easily.

All that adding the extra hedge has done is make our journey a bit longer.

Exercise. *Show that if we move the centre B to the point B^* at the end of the new hedge we can still reach it by this method.*

We can now continue this argument again. Let's build a new hedge – stuck on to the recently built one at the points E and F.

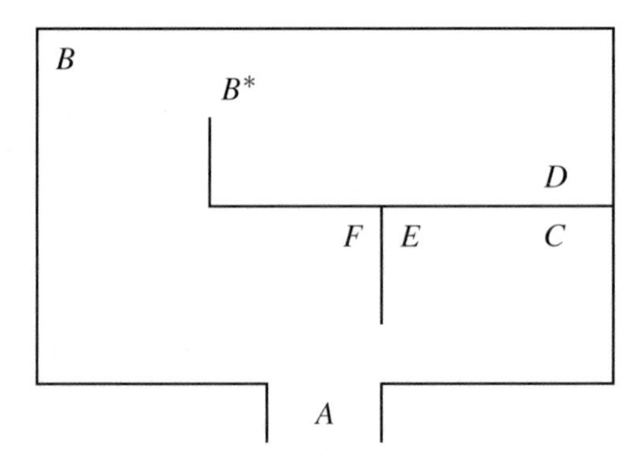

What happens now? Well we simply follow the path that cracked the maze we have just solved, and then keeping our hand on the (brand) new hedge we walk round it starting from E till we reach the point F which is (Aha!) on our original path. Then we are back to a problem we know that we can solve. Fantastic!

See if you can continue this argument yourselves by building more hedges, but always attached to a hedge that has already been built.

Exercise. *Try drawing some mazes to illustrate this.*

Provided that every hedge that we build is *always attached to an earlier hedge* we can *always get back on to our original path* and eventually get to the centre. Furthermore the 'centre' of the maze can be at any point on one of the new hedges and we can still reach it.

You should be able to show that:

> If a sparrow could walk on a hedge all the way from the finish to the start, then you can solve the maze by always keeping your hand on the hedge.

If you are in possession of a map of a maze then this gives you an easy way to test if you could solve it by the *hand on the hedge* method. In the problems there are several maps of mazes. Which can you solve using this method? This gives an easy way to test (if you are in possession of the map of a maze) which mazes can be solved using the hand on the wall trick.

1.6 How to solve a maze with a little network topology

Don't be put off by the scary words in the title of this section. Persevere and you will find a method which will solve any maze, provided that you have with you a packet of peanuts and a bag of crisps. Unfortunately, the hand on the hedge method doesn't always get us into the centre of the maze – sometimes the sparrow would have to fly to get from the finish to the start. Such a maze is shown in Figure 1.10.

The centre of the maze marked 'M' is surrounded by an 'island' which is emphasized and not connected to the outside hedge. There is no way for our sparrow to reach the outside. All that will happen with the hand on the hedge solver is that they will find themselves back at the entrance of the maze, without ever reaching the centre.

Exercise. *Try using the hand on the hedge method to show that it does not solve this maze.*

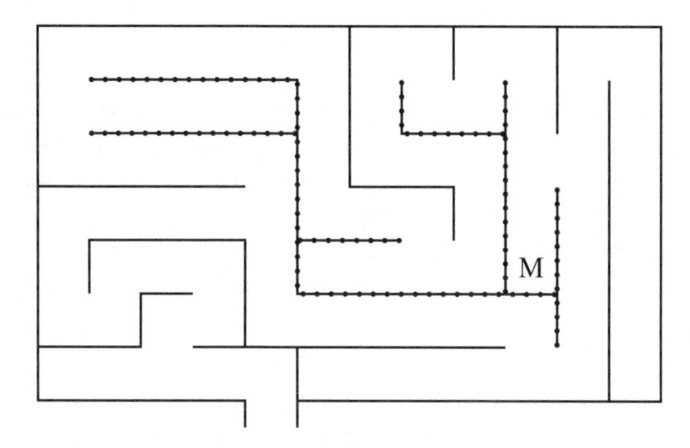

Figure 1.10: A maze with an island.

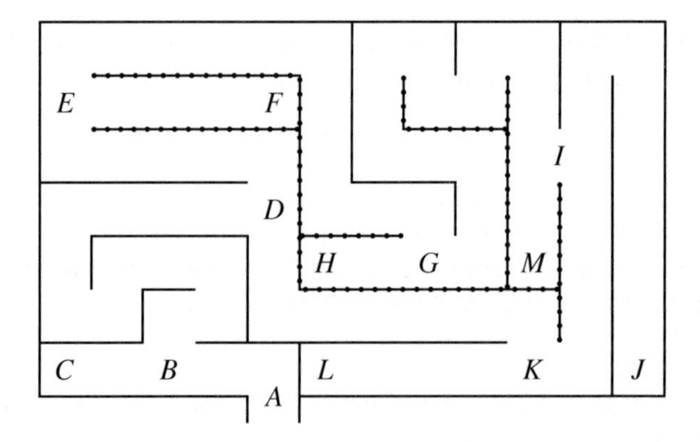

Figure 1.11: The decision points in our maze.

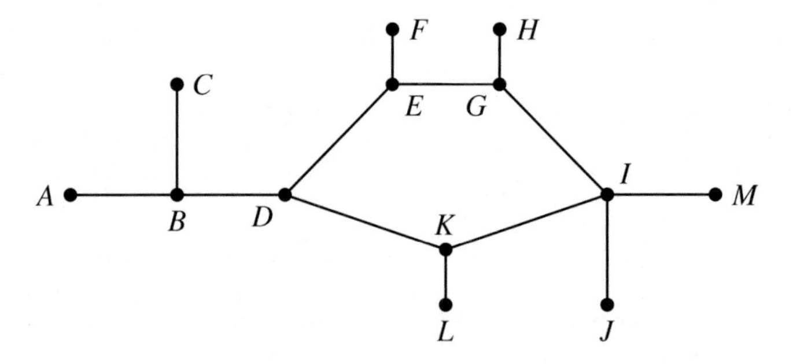

Figure 1.12: The network for our maze.

A more subtle procedure has to be used to find the way to the centre in these more sophisticated mazes. This is where *network topology* comes in: this is a way of simplifying the maze to its essential components.

When you go round a maze it doesn't matter how much you twist and turn, all that does matter are points where you have to make decisions. Life is essentially a similar process. Once you are in a bath you don't have to make any real decisions till the time comes for you to get out again. In a Cretan Labyrinth you don't have to decide anything at all, as once you enter the labyrinth you simply keep walking until you reach the centre.

In the maze in Figure 1.10 there are many points where you make decisions and these are shown as $A, \ldots, M$ in Figure 1.11. We've put a decision point at the start as you can always decide not to enter the maze. To simplify the maze we draw a network. In this network we write down all of the decision points as points on a piece of paper. We now draw paths from each of these points to the others, but only if you can go from one to another in the maze without having to make a decision in between. This gives you a *network*, shown in Figure 1.12.

Exercise. *Show that the network of the Cretan Labyrinth gives the figure below, where the only decision points are at the beginning and the end.*

Using the network it is *much* easier to see how to solve the maze, that is to find a route from the entrance A to the centre M. Indeed, we can label the solution just in terms of the decision points that you go through. For the Cretan Labyrinth this gives $A \to B$. For the more complicated maze one route to the centre is $A \to B \to D \to K \to I \to M$.

Exercise. *You should be able to find many more routes to the centre using this diagram. How many do you think there are?*

The trick of reducing a maze to its bare essentials, by finding a diagram which contains all of the information in the maze, is widely used in mathematics. A good example of this is the map of an underground railway system or metro. Often these maps only show the railway lines, stations, and interconnections. Distances between stations on the ground do not always correspond to distances on a map. But do travellers on the train care? Perhaps not as they often only need to know how to get between stations. In fact, for many people stripping away the unnecessary information might actually help then navigate successfully.

In Figure 1.13 we give a reduced map of the London Underground which was one of a series of mathematical posters which appeared on the Underground during the year 2000. This poster shows the Circle line with various connections between the stations, and it clearly resembles the network of a maze.

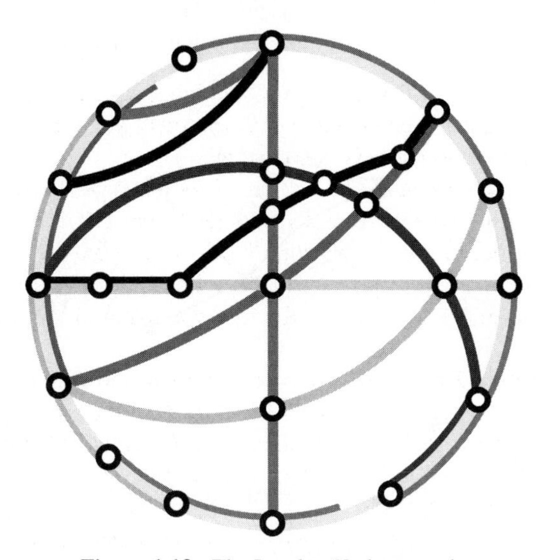

Figure 1.13: The London Underground.

Exercise. *King's Cross station is at the top. Identify the other stations and lines from the way they are interconnected.*

When we think in terms of networks, when we have a start point and finish point, then the problem of solving a maze becomes the following

> *Can you find a route in the network which takes you from the beginning to the centre and then back again?*

It is worth saying that there are really two different problems here. One is to find the route when you *don't* have a map of the maze to hand. This is the case in most recreational mazes. The second is to find the route when you *do* have a map. This case would arise if you were (for example) trying to find your way around a road network or a telephone exchange (or indeed the Underground).

For the purposes of this chapter we will consider the case when we don't have a map available.

In a network the decision points are called *nodes* and the lines connecting nodes are called *edges* or *paths*. Given a map of a network, the spaces left between the edges and the space outside are called the *faces*. If an odd number of paths meet at a node then it is called an *odd node* and if an even number of paths meet then it is called an *even node*. A dead end (such as the decision point at C) is an odd node as only *one* path leads into it. This is all shown in Figure 1.14.

Networks were first studied by the great Swiss mathematician Leonard Euler. Euler was one of the most productive mathematicians who ever lived and he created a lot of modern mathematics. Perhaps the most surprising result in the whole of mathematics is due to Euler, and can be found in Chapter 4 on magical mathematics. In 1736 Euler

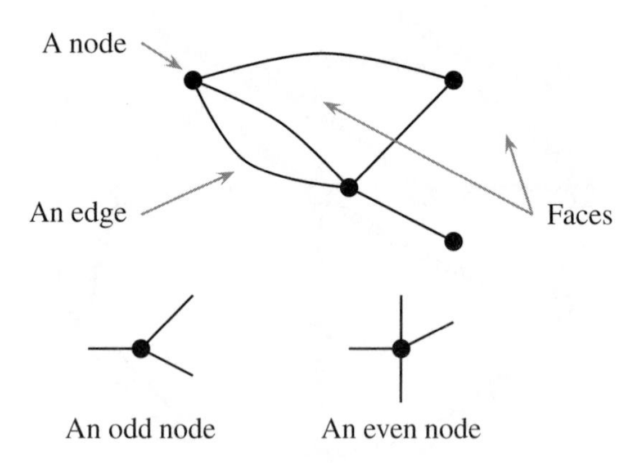

Figure 1.14: Parts of a network.

became interested in networks through trying to solve the problem of the *Bridges of Königsberg*. Königsberg, now called Kaliningrad, is a town in Russia on the Baltic Sea which has the river Pregel running through it with the island of Kneiphof in the middle of the river. The mainland and the island were connected by bridges in the arrangement shown in Figure 1.15.

The citizens of Königsberg had noticed that there seemed to be no way of going for a walk in which each bridge was crossed once and once only, but wondered whether they were being stupid and that there might be a route if only they looked hard enough. Euler took up this challenge and started by reducing the problem to a network. In this network, the nodes were the four land masses A, B, C, D and the edges were the bridges. The resulting network is shown in Figure 1.16.

The problem of the bridges of Königsberg can now be stated as follows:

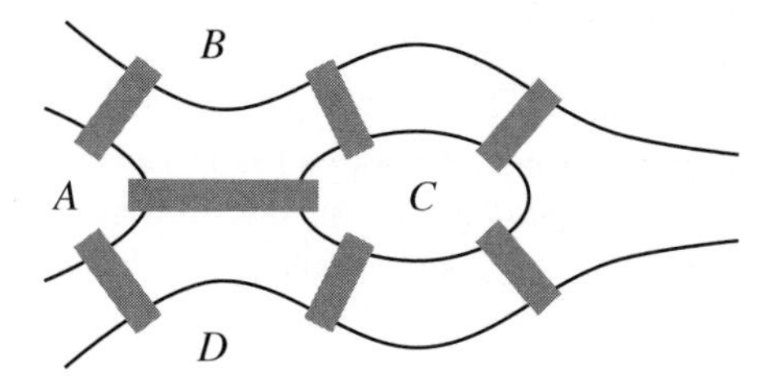

Figure 1.15: The bridges of Königsberg.

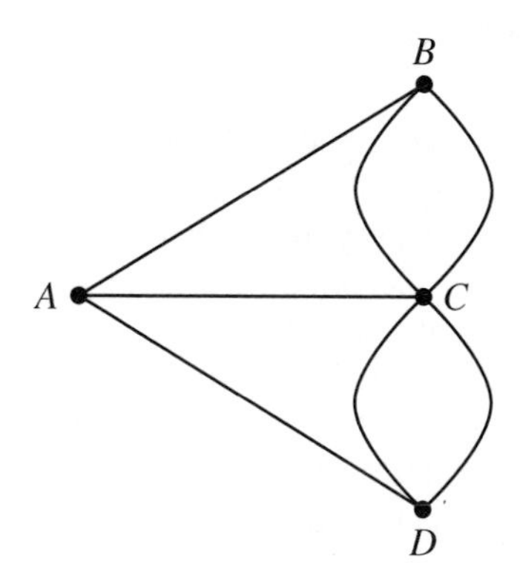

Figure 1.16: The network of the bridges of Königsberg.

> *Can you start from any node and construct a route around the network which will bring you back to the starting node and go down each edge once and only once?*

It is possible to ask this question for *any* network, not just for the one above, and Euler came up with a brilliant solution to general problem. Here it is:

1. If you have any network which has only *even* nodes then you can start at any node and find a route which returns you to that node which goes down each path once and once only.
2. If the network has exactly *two* odd nodes then you can construct a route which starts at one odd node and ends up at the other and goes through every path once and once only.
3. If the network has more than two odd nodes then there is no route that goes through every path once and once only.

It is worth pointing out that no network can have only one (or indeed any odd number) of odd nodes. See if you can do this question in the exercises. If you are interested in finding out more, then have a look at the discussion of this problem in *Mathematical Recreations and Essays* by Rouse Ball and Coxeter (listed at the end of this chapter).

The network for the bridges of Königsberg has four odd nodes so no route is possible which crosses over every bridge once and once only. To make this possible the simplest solution is to demolish one of the bridges landing on A, for example, the bridge between A and C.

Exercise. *Check that if you do this then at least you can walk from B to D going over each bridge once and only once (although you can't do this if you want to start and finish in the same place).*

This is a neat solution mathematically, although not a great idea if you happen to live in Königsberg. See the Königsberg question in the exercises.

We seem to have come a long way from solving a maze, but in fact we have nearly finished. The proof of Euler's theorem actually gives us a way of solving the maze. What we do is use the methods described in the proof to construct a route into the centre of the maze and back out again which goes down each path at most twice. To start we first take the network for the maze. In the exercises you can try constructing the network for various popular mazes. Now, these networks have a collection of odd and even nodes and this makes it awkward to use any of the results of the above theorem. Our first step is to convert the maze into one with only even nodes. This we do by the simple process of drawing each path between two nodes *twice*. What this means on the ground is that in following our way around the maze we are allowed

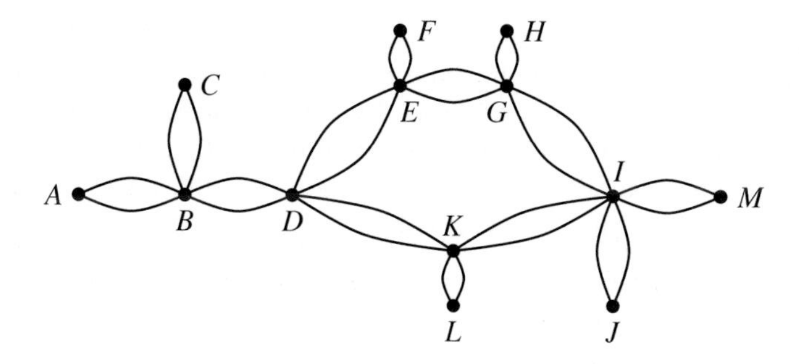

Figure 1.17: Doubling up the maze.

to take each path twice but no more. Think about this – this is very necessary. If we could only go down each path once then there would be no way out of a dead end! For our example maze in Figure 1.10 this gives the network shown in Figure 1.17.

Doubling up the number of paths in the network corresponding to our maze has converted it into a network with only even nodes. Euler's first result states that in such a network we can construct a path from *any node* which will return us to that node and which goes down each path once and once only. Now, suppose that we start at the entrance to the maze at point *A* and find this route. As it goes down every path, sooner or later it will go down a path which leads to the centre of the maze. This is a splendid start. Now continuing along our route we will eventually get back to the start of the maze again. It looks as though we have found a fool-proof way of cracking the maze. What is the catch? Well there isn't one really, except that the route we construct may not be optimal (i.e. it may be much longer than the shortest route into the centre and back). This makes the method inefficient for solving problems concerned with traffic flow (for which there are much better methods around) but it doesn't matter too much for the networks corresponding to mazes.

As we have not given the proof of Euler's theorem, we can't immediately jump to a solution. Fortunately this has already been done for us and we will describe the method of M. Trémaux, which is described in the book *Mathematical Recreations and Essays* by Rouse Ball and Coxeter. What is useful about the method we are going to describe is that you don't need to use a map of the maze, but you do need to use a packet of peanuts and a bag of crisps.

1.7 How to solve a maze using a packet of peanuts and a bag of crisps

You enter the maze, which we will assume has high hedges which you can't see over, and we will assume that all paths and nodes (where you make decisions) look very much the same. The peanuts and crisps are used as *markers* in the maze. Trail the peanuts (here and there) as you go and leave a peanut at all decision points. This will tell you whether you have been to a decision point before or whether you have gone down a path before. If you go down a path a second time, then trail a path of crisps.

If you have a rule that it prevents you ever going down a path with crisps this will stop you going down that path again. If you reach a decision point which does not have a peanut there, then we call this a *new node*. Leave a peanut there; it now becomes an *old node*. Similarly, a path without peanuts or a path in which you are currently on and dropping peanuts for the first time is a *new path*. A path with peanuts already on it and on which you are now dropping crisps is called an *old path*. Here is now how to solve any maze:

1. Start at the entrance and take any path.
2. If at any point you come to a *new node* then leave a peanut and take any new path.
3. If you come to an *old node*, or the end of a blind alley, and you are on a *new path* then turn back along this path.
4. If you come to an old node and you are on an old path then take a new path (if such exists) or an old path otherwise.
5. Never go down a path more than twice.

If you follow this procedure then you will eventually reach the centre and then get back out again. This of course only happens if no one eats the peanuts, and here we have to hope for the best! Try it out on the examples in the exercises, for which a pencil mark will substitute for the peanuts.

Exercise. *Use this method on the network in Figure 1.12 to show that the method gives as one possible route, the route* $ABCBDEFEGIMIJIKLKDKIGHGEDBA$.

Interestingly enough, if you read the account of Harris' adventures in the Hampton Court maze, you will find that he also used a marker. Instead of a peanut, they used a baby's bun which showed them when they had come back to the same point. Unfortunately as there was only one bun available it didn't help much with solving the maze itself and left the baby hungry.

Exercise. *Can you discover why and how this method works?*

1.8 Modern mazes

The method of using crisps and peanuts for solving a maze is a good one if you don't have a map of the maze and have a lot of time. This is usually the case in a puzzle maze. However, in other mazes you may well have a map and not much time to solve it. A good example is the road traffic network, where you need to find a route quickly around the one-way system, or alternatively, you might want to find a fast route between two widely separated towns. To solve this sort of problem, computer scientists have derived some really fast methods, known as *algorithms*. One of the best is the so-called *flood* algorithm in which we imagine that we pour water into the entrance of the maze. Sooner or later some of the water will reach the centre. When this happens you can find the route in. Similar methods are used to solve other problems using 'real life' mazes, such as the problem of wiring up a complicated circuit board.

We have come a long way from the legend of the Minotaur to the design of computers. But the same ideas behind the maze apply to both, and give us a wonderful mathematical link from the past to the present.

1.9 Exercises

First session

1. Think of as many examples as you can of mazes appearing in real-life situations.
2. Design a really fantastic maze and then challenge a friend to see if they can solve it.
3. Draw a Cretan Labyrinth. If you are on a beach, draw a giant labyrinth in the sand. If you draw it big enough you can walk around it.
4. Suppose that a Cretan Labyrinth is 5 cm in diameter. How long is the route from the beginning to the centre?
5. Draw a labyrinth starting from each of the following seeds.

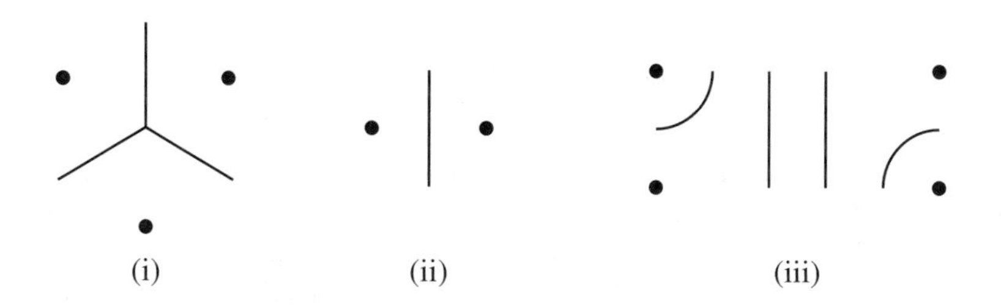

6. Now draw your own seed and see what sort of labyrinth you can make from it. Does it have an entrance (or several entrances) and a centre (or several centres)?
7. Draw a sequence of points in a line and label them

$$\ldots\ I\ G\ E\ C\ A\ B\ D\ F\ H\ \ldots$$

Draw a path in the order $A \to B \to C \to D \to \ldots$. You should end up with a spiral (which you can enlarge if necessary). In this labyrinth you have an entrance and a centre. Now consider the points labelled with the sequence

$$\ldots\ F'\ E\ D'\ C\ B'\ A\ A'\ B\ C'\ D\ E'\ F\ \ldots$$

Join up the points in the order $A \to B \to C \to D \to \ldots$ and then the points $A' \to B' \to C' \to D' \to \ldots$ working round anticlockwise. You should end up with a *double* spiral which has an entrance, a centre, and an exit. See if you can combine several such double spirals to make a really complicated labyrinth.

In fact most labyrinths boil down to double spirals or combinations of double spirals. The Cretan Labyrinth is really a double spiral bent around a circle and the Roman mosaic in Figure 1.7 is four double spirals joined together.

Second session

1. Here are some maps of some well-known mazes, see also the map of Hampton Court given earlier. In each case the centre is marked by an X. Which of these can be solved by using the *hand on the hedge* method? Check your answers by solving the maze in each of these cases.

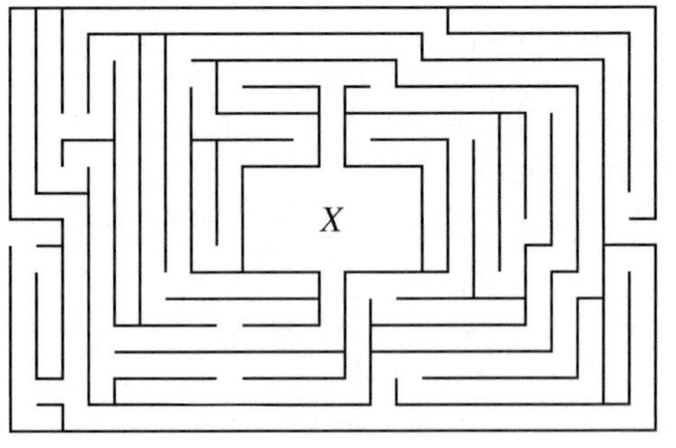

Hatfield House.

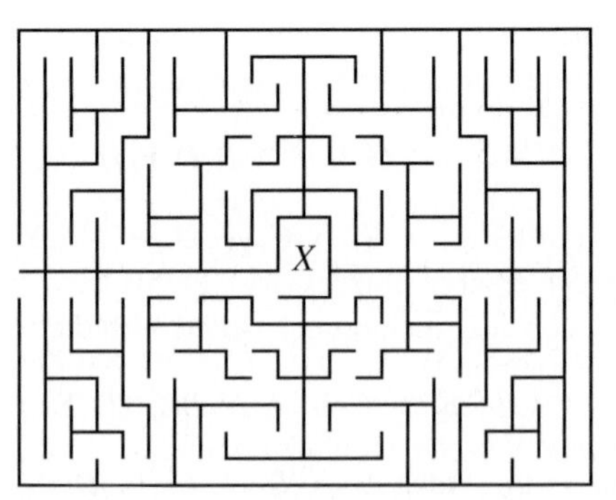

The plan of a maze in Nova Scotia.

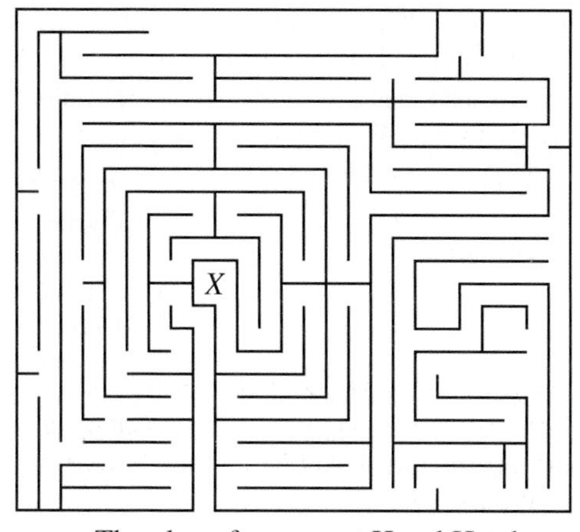

The plan of a maze at Hazel Head.

2. For the mazes in the above question, draw a network for each with nodes and paths. Using the network see if you can find the route to the centre. Try to find several such routes. Repeat this for any other networks you can find.
3. Can you see why every network must have an *even* number of *odd* nodes?
4. Solve the above mazes by using the *peanuts and crisps* method. Construct your own mazes and try the method on these.
5. As bridge builder for Königsberg you have been asked to build some extra bridges so that everyone can take a walk from any point back to that same point going over each bridge once and only once. Where would you build these bridges? What is the smallest number you need?
6. If you are in a dramatic mood, build a labyrinth in the classroom and (in appropriate costume) act out the story of Theseus and the Minotaur (don't forget Ariadne and the thread).

Field trips and projects

It is great to visit a maze and makes a splendid day out. Table 1.1 shows some mazes open to the public. A much fuller list of mazes around the world can be found in the books below.

- The AMAZING HEDGE PUZZLE at Symonds Yat West, Ross-on-Wye, Herefordshire HR9 6DA, England. Run by Lindsay and Edward Heyes, this site includes the Jubilee Maze, the Museum of Mazes, a restaurant, a puzzle shop, and activity centre. The Heyes brothers will (if asked) lay on demonstrations about how to build a maze, and will take you for games of pursuit and evasion inside the maze (Figure 1.18). Details as well as a whole load of extra information on mazes can be found on `http://www.mazes.co.uk/`

Table 1.1: Some mazes around the world

Country	Town and County	Location
Belgium	Bruges	Chateau de Loppem, near Bruges
Czechoslovakia	Prague	Prague Mirror Maze, Petrin Hill, Prague
France	Indre-et-Loire	Chateau de Chenonceau, Indre-et-Loire
France	Paris	Jardin des Plantes
France	Ponce, Loire	Chateau de Ponce, Ponce
New Zealand	Rotorua, North Island	Rotorua Hedge Maze, Rotorua
United Kingdom	Blackpool, Lancashire	Blackpool Pleasure Beach
United Kingdom	Castle Bromwich, West Midlands	Castle Bromwich Hall Gardens Maze
United Kingdom	County Antrim, Northern Ireland	Carnfunock Amenity Park, near Larne
United Kingdom	Crystal Palace, South London	Crystal Palace Maze
United Kingdom	East Molesey, Surrey	Hampton Court Palace
United Kingdom	Edenbridge, Kent	Hever Castle
United Kingdom	Edinburgh, Scotland	Edinburgh Zoo
United Kingdom	Falmouth, Cornwall	Glendurgan House, near Falmouth OS ref. SW 772277
United Kingdom	Hay-on-Wye	Llangoed Hall – Llangoed Hall, near Hay-on-Wye
United Kingdom	Isle of Wight	Blackgang Chine
United Kingdom	Maidstone, Kent	Leeds Castle Maze, Leeds Castle, near Maidstone
United Kingdom	Romsey, Hampshire	The Clock Maze, Paultons Park, Ower, near Romsey
United Kingdom	Ross-on-Wye, Hereford & Worcester	The Jubilee Maze and Museum of Mazes, Symonds Yat West, near Ross-on-Wye
United Kingdom	Saffron Walden, Essex	The Hedge Maze, Bridge End Gardens
United Kingdom	Scarborough, North Yorkshire	Victoria Park Maze, Victoria Park
United Kingdom	St Johns, Isle of Man	Hedge maze in Forestry Commission plantation
United Kingdom	Woodstock, Oxfordshire	Blenheim Palace
United Kingdom	Ynys Mon, Cymru	Wylfa Nuclear Power Station
USA	Brekenridge, Colorado	The Brekenridge Maze, 710 South Main Street, Brekenridge
USA	East Tawas, Michigan	Maze of the Planets, East Tawas
USA	Florida	Daytona Beach
USA	Great Gorge, New Jersey	The Puzzle Place, Vernon Valley Action Park
USA	Williamsburg, Virginia	The Williamsburg Maze, The Governor's Palace

Figure 1.18: A field trip in action: constructing a Cretan Labyrinth at the Amazing Hedge Puzzle, Symonds Yat.

- LONGLEAT, Wiltshire. Not for the faint-hearted, this is advertised as being the World's largest maze. It is certainly a very difficult maze to solve. If you go with small children, hold on to them tightly. Nearby are smaller mazes such as the Lunar Labyrinth, the Sun Maze, and the Maze of Love, as well as a safari park. Details can be found at `http://www.longleat.co.uk`.
- HAMPTON COURT. `http://www.hrp.org.uk/`.

Other projects could involve finding out more about how information packets are transmitted over the internet.

1.10 Further problems

Strictly speaking a network is completely defined when we list the nodes and the connections between them. Such lists aren't always that easy to interpret. People are usually much better at interpreting pictures. We are going to draw pictures of networks but when representing networks in the following exercises we must make sure that the network diagrams all satisfy the following ground rules. Go back and look at Figure 1.14 if you need to be reminded of the names of the parts.

The network diagram ground rules

- The network is connected

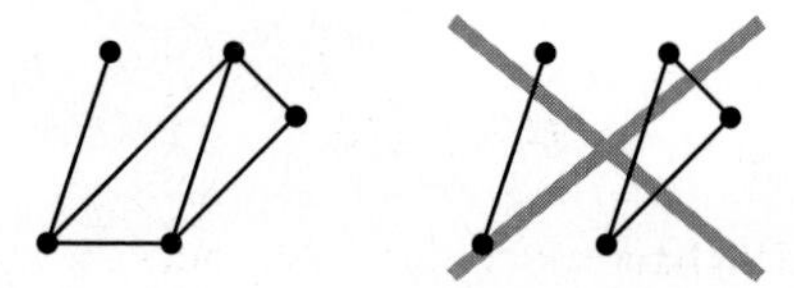

- Edges do not cross

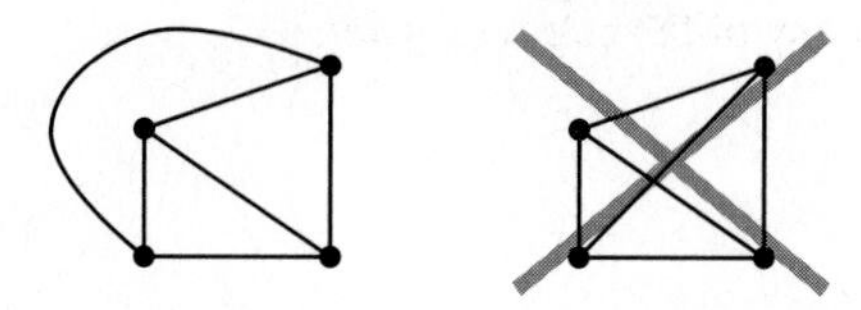

- The lengths of the edges are unimportant

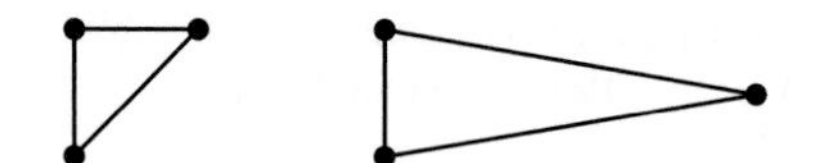

These pictures represent the same network.

- The positions of the nodes are unimportant

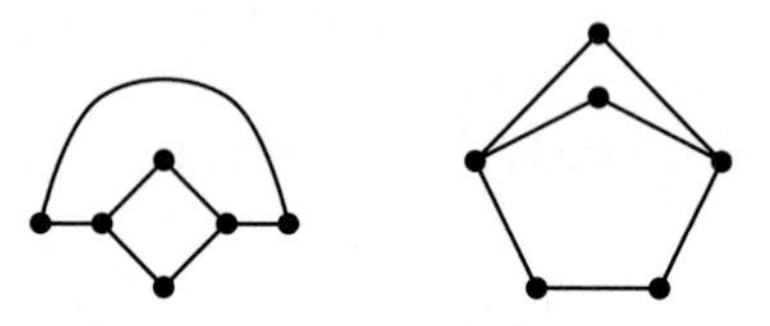

These pictures represent the same network.

From these rules it is clear that each network has many pictures.

Problems

1. Draw at least five network diagrams of your own. Count the number of nodes, edges, and faces and fill in a table with your answers.
2. For each network diagram count the number of nodes, edges, and faces, and calculate the value of

$$\text{nodes} - \text{edges} + \text{faces}.$$

What do you notice? This is known as *Euler's formula* and it is very famous.

3. Can you redraw the following network diagram without any of the edges crossing?

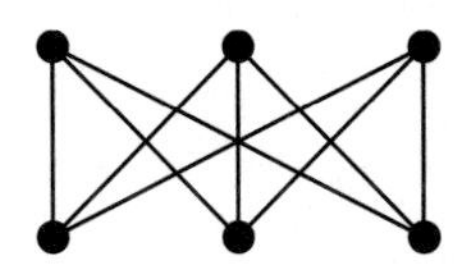

4. Make a squashed doughnut by cutting a big hole in a paper disk. Draw some networks on the doughnut. Use both sides and make sure some edges go round the outside and through the middle. Calculate the value of Euler's formula.
5. Can you draw the above network diagram from question 3 on the surface of a doughnut without any of the edges crossing?

1.11 Answers

Since many of the questions require you to draw mazes of your own, etc., not all of them have answers below.

First session

7. Here is a double spiral

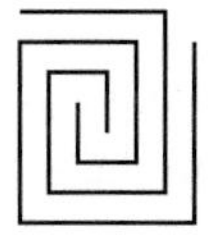

Now combine copies to produce a more complex labyrinth such as

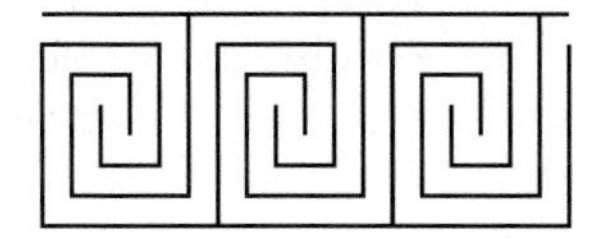

Second session

3. Every network has an even number of odd nodes. This is called the *hand-shaking principle* by some mathematicians. Each edge in the network has two ends, each of which is connected to a node. If there were an odd number of odd nodes there would be an edge from one node that didn't go anywhere.
5. One solution would be to double up every bridge but we can do better than that. Euler's rules state that we need to end up with *no odd nodes*. We have four odd nodes so we can add two edges in many ways so that we have no odd nodes left.

Adding only one edge will not be enough. An example would be to build a bridge between B and D and one between A and C but there are many other combinations.

Further problems

2. Euler's Formula tells us that

$$\text{nodes} - \text{edges} + \text{faces} = 2.$$

If you don't count the outside as a face then the total will be 1. You can find out more about this important formula from the books in the references.

3. This network is known as $K_{3,3}$ and is one of the simplest network diagrams that can't be drawn on a sheet of paper without edges that cross.
4. The value of Euler's formula for networks on the surface of a doughnut will be different. Some networks will give a value of 2 as before. Some, however, will give a value of 0.
5. The network diagram for $K_{3,3}$ can be drawn on the surface of a doughnut without any edges crossing.

1.12 Mathematical notes

Mathematicians often call networks *graphs* and so the subject is known as *graph theory*. There are many problems associated with graphs and they are important both to pure and applied mathematics.

As hinted at in the further problems the value of

$$\text{nodes} - \text{edges} + \text{faces}$$

tells us as much about the surface we are writing on as the network we are drawing. In fact this formula is used to classify different surfaces. For doughnut-like surfaces

$$\text{nodes} - \text{edges} + \text{faces} = 0.$$

A discussion is given in the book by David W. Farmer and Theodore B. Stanford.

1.13 References

Books

There are many good books on mazes, and also lots of shops stock maze puzzle books which are fun to solve. Here are some books that we like

- Rouse Ball, W. W., and Coxeter, H. S. M. (1987). *Mathematical Recreations and Essays,* Chapter IX. Dover.

A wonderful book that we will refer to often throughout this text.

The following books are all good sources of further information about mazes.

- Field, R. (1999). *Mazes, Ancient and Modern.* Tarquin.
- Matthews, W.H. (1970). *Mazes and Labyrinths: Their History and Development.* Dover.
- Bord, J. (1976). *Mazes and Labyrinths of the World.* Latimer.
- Fisher, A. (1991). *Mazes.* Shire.
  As well as containing a lot of information about mazes, the author also maintains a web-site `http://www.mazemaker.com/` with a feast of information about mazes around the world.
- Jerome, J. K. (1953). *Three Men in a Boat (to say nothing about the dog).* Pitman.
  Not a book about mazes, but a wonderful account of the maze at Hampton Court together with much other amusing material.
  More information about the mathematics of networks and the Euler formula can be found in:
- Farmer, D. W., and Stanford, T. B. (1995). *Knots and Surfaces: A Guide to Discovering Mathematics,* Mathematical World, Vol. 6. American Mathematical Society.
- Wilson, R. J. (1966). *Introduction to Graph Theory.* Longman.

2
Dancing with mathematics

2.1 Introduction

In the popular imagination, the mention of folk dancing conjures up many images. Brownies shuffling round an old church hall, while Brown Owl plays the piano and Tawny Owl sings encouragement. Morris dancers, decked out in flowers and ribbons and only slightly drunk, hitting each other on the head with large sticks. Mad Scotsmen, hurtling across the mountains, jumping on swords, and doing strange things with a haggis. This all seems rather remote from mathematical equations like $a^2+b^2=c^2$. So why should mathematicians be at all interested in folk dancing? One possible explanation is that to the general public, mathematicians and folk dancers have one thing in common: they are eccentric, mad, and generally detached from reality. (Indeed it was Bernard Shaw who said that you should try anything once, apart from folk dancing.)

But there is a much more significant connection between mathematics and folk dancing: both are concerned with *patterns*. To be precise we can think of a folk dance as the performance of a series of simple motions according to a set of rules. The purpose of these motions may seem strange at first but when combined together they can produce some wonderfully complex dances involving every girl or boy in the room. Indeed some of the best dances involve very complex patterns which marvellously simplify at the end of the dance so that you always end up with the same partner as the one you started with. If you watched such a dance from a bird's eye view, you would see each dancer trace out an intricate path carefully interwoven with those of the other dancers. There in a nutshell is the basis for nearly all of the well-known traditional English, Scottish, and American dances such as Lucky Seven, Dashing White Sergeant, Nottingham Swing, Dorset Four Hand Reel, most square dances and, of course, the Hokey-Cokey (always a good dance to end an evening with, even if the mathematics behind it is a little simplistic). We shall show in this chapter that the patterns formed by the men and women in such dances are closely related to shapes like triangles and squares. Furthermore, the basic mathematics behind the patterns in a folk dance can also be applied to generate patterns in quite different areas, such as making up the tunes in bell-ringing, or knitting jumpers. This demonstrates the real power of good mathematics, that is the ability to transfer ideas rapidly from one area to a quite different one.

2.2 Symmetry and group theory

The basic mathematics behind folk dancing is group theory and in this section we will give you a small flavour of this beautiful branch of mathematics. Group theory was pioneered in the early nineteenth century by the French mathematician Evariste Galois, who subsequently died in a duel at the age of twenty one, and is generally regarded as one of the more *romantic* mathematicians in history. Group theory is usually first encountered by undergraduates (although, sadly, its links with folk dancing are not always stressed). As well as being a beautiful and significant branch of mathematics in its own right, it also has important applications in physics (in particular the classification of elementary particles), chemistry (where it is used to help in the study of crystals), and in cryptography (see the chapter on codes and ciphers). Group theory can be described as the mathematics behind symmetry and patterns.

Before the next dance it is worth having a short interval, drinking some lemonade, chatting to your partner and, in this chapter, looking at what we mean by symmetry. Symmetry is all around us, in biology and physics, in art and in music. A snow-flake is an excellent example of a beautiful natural shape which is symmetric (indeed it has a six-fold symmetry), as is a virus, a shell, a starfish, or the spots on a leopard's coat. Human beings seem to appreciate symmetry instinctively and much great art is based upon it, examples being stained glass windows, the pictures of Escher, and the music of Bach. Symmetry even finds its way into poetry, and here is an important example by the poet William Blake.

Tiger Tiger burning bright
In the forests of the night
What immortal hand or eye
Could frame thy fearful symmetry

So, what do we mean by symmetry? We will start our look at symmetry by considering a concrete object like a square. For the moment we think of a square as a geometrical object. Later on by making some abstractions we can turn what we learn into a square dance. (The pun is only half intentional.) Here in Figure 2.1 is an example of a square with the four corners labelled A, B, C, D. These labels will come in useful later for the dancing. The labelling used here is necessary for the dances we will describe. Later on we will see how a different labelling gives a different dance.

We can perform various geometrical operations on the square which change the positions of its corners but do not actually change its shape or position. These operations are called the *symmetries of the square*. A little thought will tell you that these operations are either rotations through whole multiples of 90° or they are reflections through the centre-lines of the square or through its diagonals. Here, for example, in Figure 2.2 we show the effect on the square of a clockwise rotation about its centre by 90°. In this operation all of the corners of the square change their positions.

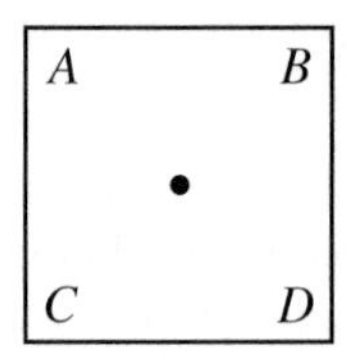

Figure 2.1: The square in its staring position.

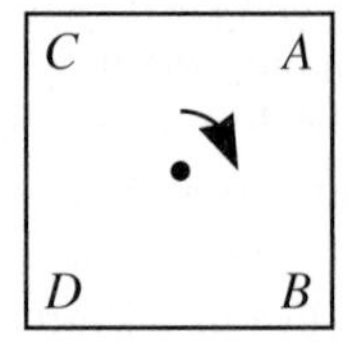

Figure 2.2: The square after a rotation of 90° clockwise.

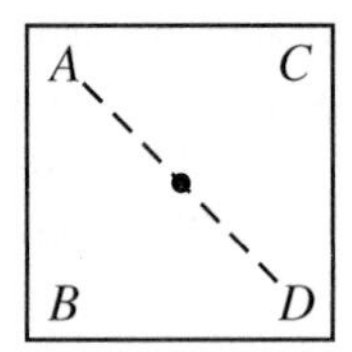

Figure 2.3: The square after a reflection through the diagonal.

Exercise. *You will find the following pages* much *easier to follow if you pause for a moment and find a square of paper. Cut one out carefully if you need to and mark the four corners with A, B, C, and D. You can use this square to help you understand the patterns we are about to make.*

In Figure 2.3 we show the effect of a reflection of the square illustrated in Figure 2.1 about the diagonal line which passes through the corners presently at A and D and which is at an angle of $-45°$ to the vertical. In this operation the corners A and D stay where they are and the corners B and C swap places.

Figure 2.4 shows a reflection of the square illustrated in Figure 2.1 about the vertical line through the centre which bisects the lines AB and CD.

Exercise. *Show that if we count the operation of leaving the square unchanged as a rotation of* $0°$*, then the square has eight symmetries, of which four are rotations and four are reflections.*

To save having to produce any more pictures of the square we can introduce a short-hand to represent these operations.

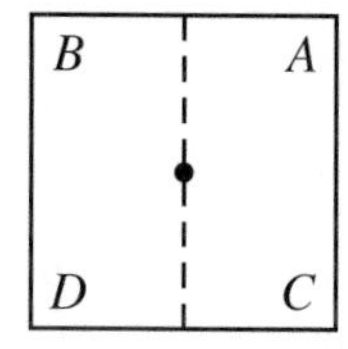

Figure 2.4: The square after a reflection through a vertical line.

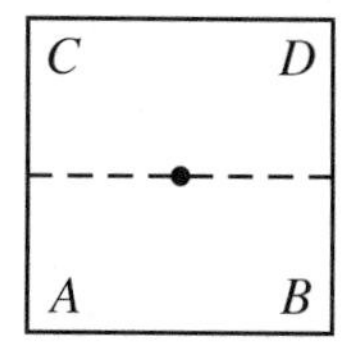

Figure 2.5: The sequence after a reflection through a horizontal line.

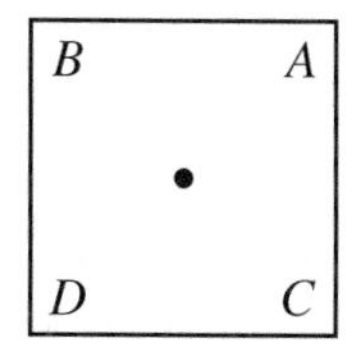

Figure 2.6

In this short-hand the 90° clockwise rotation illustrated in Figure 2.2 will be represented by the symbol a, the reflection through the diagonal illustrated in Figure 2.3 by b, and the reflection about the vertical illustrated in Figure 2.4 by c. We express the effect of combining *two* operations on the square by writing down sequences of these letters. We will use the following convention. The operation ab will be the combined operation on the square of first rotating it clockwise through 90° and then taking the resulting square and reflecting it through the diagonal at $-45°$ to the vertical. If we do this we get the square shown in Figure 2.5.

Have a look at this. The combination of the two operations is in fact another reflection – this time about a horizontal line. We can call this operation d so that

$$ab = d.$$

Now let's try the same operations but in the opposite order, to give an operation ba which represents a reflection of Figure 2.1 about the diagonal followed by a rotation. If we do this then the arrangement of the square is illustrated in Figure 2.6.

We have seen this before: it is the operation of reflection about the vertical line which we called c, so that

$$ba = c.$$

Something rather remarkable has happened here. We have shown that $ab = d$ and $ba = c$. But c and d are different operations, so the *order* in which we perform our operations is vital. This is quite unlike the operations of addition or multiplication which do not depend upon the order in which you do them.

In this representation aa will be two rotations (i.e. a rotation through 180°) and aaa a rotation through 270°. Now what is $aaaa$? This is a rotation through 360°, which is the same as a rotation through 0°, which is the operation which leaves the square unchanged. We call such an operation the *identity* operation e so that

$$aaaa = e.$$

The identity operation has the property that if you combine it with any other operation you always get the same operation so that

$$ae = ea = a \quad \text{and} \quad eb = be = b.$$

Exercise. *There are other ways of combining operations to give the identity. Convince yourselves that any reflection if repeated by the same reflection gives the identity, so that*

$$aaaa = bb = cc = dd = e.$$

Here is an example of how operations can combine. What simple operations correspond to bc and to dc?

Now, we have already shown that $c = ba$ and $d = ab$.

Exercise. *Now see if you can follow the following calculation*

$$bc = bba = ea = a \quad \textit{and} \quad dc = abba = aea = aa.$$

In English this says that a reflection through a diagonal line followed by a reflection about a vertical line has the same result as one rotation of 90°; and a reflection through a horizontal line followed by a reflection through a vertical line has the same result as a rotation through 180°.

The set of all of the operations we have described on the square is called a *group*. This is a very special example of a group and in fact there are many other types of group around. We have seen that the operations a, b, c, etc., combine in a special way. Now suppose instead that we have a general collection of operations (which need not be related to any geometrical object). For such a set of operations to form a group they must have four properties:

1. The identity operation e must be in the set.
2. Any two operations (say a and b) when combined like ab must form another member of the set.
3. If a is an arbitrary member of the set, then there must be another *inverse* operation (called a^{-1}), also in the set, such that $a^{-1}a = aa^{-1} = e$.
4. If a, b and c are in the set, then $(ab)c = a(bc)$ (so that the effect of combining a with b and *then* with c is the same as a combined with the operation bc).

In the exercises you can check that the operations on the square really form a group.

There are very many examples of different groups and mathematicians call the study of them *group theory*. Some other examples are: the group of reflections and rotations of the triangle, the group of reflections and rotations of the cube, and the group of permutations (rearrangements) of three items. Many groups have special names, and the group of the rotations and reflections of the square is called D_4 or the *dihedral group* on four elements. We will meet other examples of groups in Chapter 6 on codes and ciphers.

Although there are an infinite number of different possible groups, a remarkable result, proved in the early 1980s, gave a classification of all groups with a finite number of operations. Details of this classification scheme can be looked up in the *Atlas of finite groups*.

2.3 Back to dancing

We are now going to translate the abstract theory of the last section into some folk dances. We will start by considering a dance with four dancers called (conveniently) Andrew, Bryony, Chris, and Daphne, who we will abbreviate to A, B, C, D.

To relate the square considered in the last section to people we will identify the dancers A, B, C, D with the corners of the square in the order shown in Figure 2.1. This **does not** mean that the dancers need to stand at the corners of a square. In fact we will think of them standing in line like

$$ABCD.$$

The important thing is that the order in which they stand in this line is the same as the order in which the corners appear on the square illustrated in Figure 2.1.

Now suppose that we apply reflection b to the square to give the picture in Figure 2.3. If we identify the new corners in this figure with the new positions of the dancers we get the arrangement

$$ACBD.$$

What has happened on the dance floor is that the two 'inner dancers' B and C have changed places and the two 'outer dancers' A and D have stayed in the same place.

In some dances they can do this by holding hands, and in others by brushing shoulders, whatever suits the music. We will call this dance move an *inner twiddle*.

Notice that we have connected the rotation operation b on the square to a rearrangement (or a dance move) of the four dancers. It is important to note that, mathematically, these two operations are completely identical, even though they appear to the eye to be rather different.

Applying b twice combines two inner twiddles, and brings the line back to its original position, just as two reflections return the square back to its original position. This is another common dance move, popular in square dancing where it is called a *Dos-e-Dos*, which is French for back to back, describing how the move is often performed.

Now let's see what happens to the line if we apply the reflection c to the square. After looking at the position of the corners in the square in Figure 2.4 we see that this corresponds to the movement

$$ABCD \to BADC.$$

What has happened here is that the two outer dancers have changed places with their two neighbouring inner dancers. For obvious reasons we will call this an *outer twiddle*.

The inner twiddle and the outer twiddle are simple moves on their own, but we can combine them to give a much more complicated (and satisfying) dance move.

We showed earlier that

$$bc = a \quad \text{and} \quad aaaa = e.$$

If we combine these two results we can see that

$$\boxed{bcbcbcbc = aaaa = e.}$$

This is a piece of mathematics; what does it mean on the dance floor? Well, the operation bc corresponds to an inner twiddle followed by an outer twiddle. This takes us from $ABCD$ to $CADB$. Mathematically this *must be* a 90° rotation of the square.

Exercise. *Check that $CADB$ really does correspond to the arrangement of the corners in Figure 2.2*

Now, what the calculation in the above box tells us is that if we do the operation of an inner twiddle followed by an outer twiddle *four times* then we (very satisfyingly) get back to the original positions. In the process of doing this, an individual dancer will go through every position twice before getting back to the start (with his or her original partner). The complete sequence is shown in Figure 2.7.

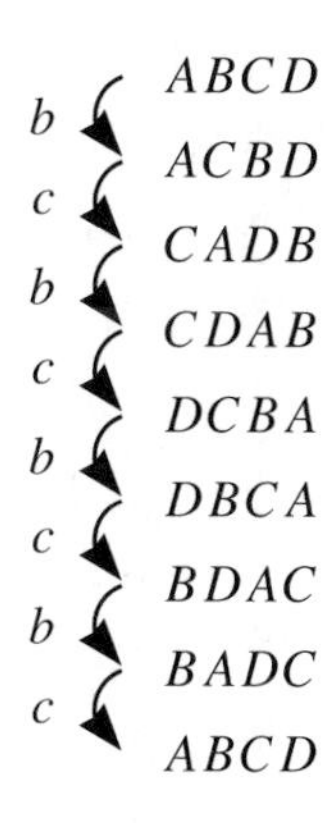

Figure 2.7

Exercise. *Try the move with four of you in a line. It is very effective if you make large letters and hold these up whilst you do the moves as then you can see exactly how the rearrangements are working.*

In English folk dancing this move is called a 'Hay', in Scotland a 'Foursome Reel', and in square dancing a 'Reel of four', but it's all the same mathematics. We will call it a reel of four from now on. A whole dance called a 'Dorset four-hand reel' is based on four reels of four, together with some stamping moves called 'ranting' which are usually done wearing clogs. It can be fun to hold hands as you cross over in each of the twiddle moves.

It is worth digressing for a moment and considering how these moves are performed. In much of English, Scottish, and American folk dancing, the music for the performance is performed in eight-bar groups, often with four such groups making a 32-bar sequence. Folk dancers in general, and square dance callers in particular, get very practised in counting in units of eight. When performing a reel of four, two options are available to you. In the quick option you take four bars over each twiddle so that the (rotation) move of an inner twiddle followed by an outer twiddle takes eight bars, and the complete reel of four then fits in to the 32-bar sequence. In practice this is often too fast for inexperienced dancers and it is easier to take eight bars over each twiddle and to then do the complete reel in two 32-bar sequences.

Armed with a little imagination, the world is now open to us to design many more dance sequences based upon the operations of symmetry, and we suggest that you have some fun making up some new ones.

For a first example, consider the reflection d. From Figure 2.5 we see that the effect that d has on the line is the rearrangement

$$ABCD \to CDAB,$$

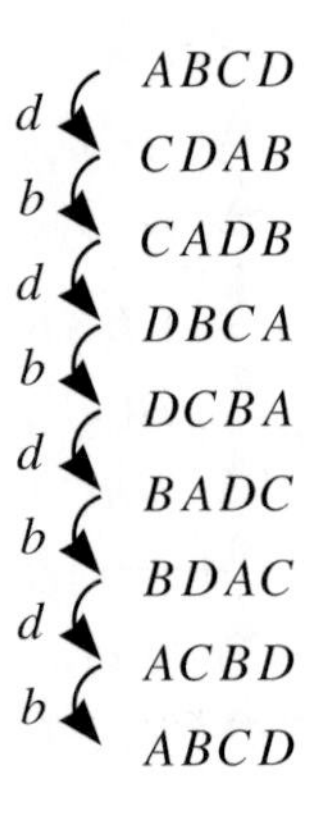

Figure 2.8

so that the couples (AB) and (CD) swap places. We will call this a *swap*. If we follow the swap by an inner twiddle then we get db. But

$$db = abb = ae = a$$

so, as before the combination of a swap and an inner twiddle equals the rotation operation a. Doing this whole process four times get us back to the start again giving another perfect dance. The resulting sequence is shown in Figure 2.8.

Of course, not all folk dances have only four dancers. Indeed, in a good square dance you may have as many as 100 dancers. To construct a dance for more people we can again apply the general principles of group theory, but this time to a sequence with more letters. As group theory is a vast subject, and not all groups correspond to good dances, it is worth saying that the best dances correspond to simple rearrangements which combine together to give complicated patterns. Here, as an example, is a simple dance for eight. Suppose that we have four boys A, B, C, D with partners E, F, G, H who stand opposite thus

A	B	C	D
E	F	G	H

In this line we call AE the *top couple* and DH the *bottom couple*. We can denote this arrangement more compactly by the sequence $ABCDEFGH$.

A very common dance move is called *casting*. In the best form of this, the *top couple* AE each turn outwards and walk to the end of the line, followed by everyone else. At the bottom of the line they make an arch and everyone goes through it (holding hands) and walk back to their original places, apart from the couple AE who stay at the bottom. This probably sounds rather complicated, but the final arrangement is just

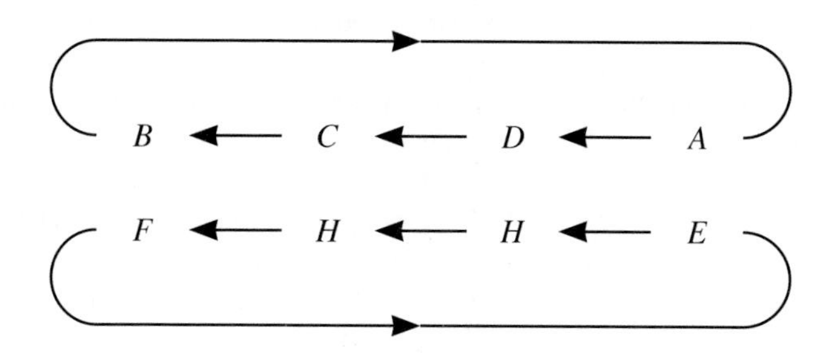

so that BF are the new top couple and AE the new bottom couple. As a rearrangement on the whole sequence this corresponds to:

$$ABCDEFGH \to BCDAFGHE.$$

Suppose that we label the operation producing this rearrangement as a.

Exercise. *Show that*

$$aaaa = e.$$

Indeed, a is rather like the rotation of the previous section. Another move called *crossing* is given by

$$ABCDEFGH \to FEHGBADC.$$

Exercise. *Have a look at what this corresponds to in terms of the actual motions of the dancers and think how you might actually do it. (Hint: in practice this is often called a star.)*

Suppose that we call the crossing (or 'star') move b.

Exercise. *Show that* $bb = e$.

We can see that b is rather like the reflection of the previous section. The sequence bba (crossing, crossing, casting) repeated four times gives

$$bbabbabbabba = aaaa = e.$$

This dance is called 'Farmer's Jig' and is often performed 'at the gallop' to the tune of *John Brown's Body*.

2.4 Bell-ringing and knitting

Flushed with our dancing success we can now turn out attention to other pursuits which involve patterns and symmetry. The symbols A, B, etc., which we have rearranged to make the dance patterns, can in fact represent any objects for which a sequence of rearrangements makes sense. A good example is to be found in bell-ringing: bells are rung in different orders to signify marriage, death, imminent invasion, etc. Suppose that $ABCD$ are four bells. Typically such bells hang in a church tower and are rung by bell-ringers pulling at ropes. (Indeed there are often more than four bells in a tower.) A sequence such as $BCAD$ would correspond to the order of pulling at the ropes to ring a *peel*. A second rearrangement such as $BCDA$ would then give another peel. A sequence of peels gives a complete order for ringing the bells which might (for example) be used just before a wedding. Diagrams such as Figures 2.7 and 2.8 then tell you in which order to 'pull the ropes' and are much used by the bell-ringers, some of whom are well aware of the underlying mathematics. It is worth noting that though mathematically similar, there are big differences between the nature of the rearrangements that you get in bell-ringing and folk dancing. Firstly, in dancing the moves usually exploit the fact that male and female dancers are different. For example if A and B are two dancers of the same sex, then many dances would not have a move which just involved A and B alone. This can reduce the number of possible rearrangements. As bells are neither male nor female there is no such restriction in bell-ringing. In contrast, in bell-ringing a rearrangement is only allowed if a bell (say A) does not change its position by more than one place at each change. Thus $ABCD \to BACD$ is possible but $ABCD \to CBAD$ is not. Thus some of the dance moves are not possible in bell-ringing.

On a more colourful note, patterns similar to Figures 2.7 and 2.8 can be used in knitting. Suppose that you have a jumper which is made up of several lines of stitches. Patterns on jumpers can be made by having the stitches of different coloured wool. We can use group theory to help us generate some nice patterns. Suppose that we take the rearrangements illustrated in Figure 2.7 and think of A and D as black wool and B and C as red wool. If we now draw lines to mark where each letter is we get a pattern. In Figure 2.9 we have placed two such patterns side by side to give a jumper pattern. In Figure 2.11 there is a photo of one of us wearing the very same jumper.

We can generate lots of different knitting patterns this way. As an example of another one we will try changing the way in which we have used the arrangement of the corners of the square. Suppose that we change the order of the corners so that $ABCD$ now corresponds to a square labelled as in Figure 2.12. This time, instead of reading off the letters as $ABCD$ as we have up till now, we will read *clockwise round the square* to give $ABCD$. You should compare this with Figure 2.1. Reading off the letters in a different way will change the patterns we get.

Let's now look at what our earlier operations do.

Figure 2.9: Knitting pattern from Figure 2.7.

Figure 2.10: Knitting pattern from Figure 2.8.

Exercise. *If we perform the reflection c on the square in Figure 2.12, show that this induces the permuation*

$$ABCD \to BADC.$$

Thus c is still an outer twiddle.

Exercise. *Now look at what happens if we apply the reflection d. Show that we now get*

$$ABCD \to DCBA.$$

Figure 2.11: A jumper with our knitting pattern on.

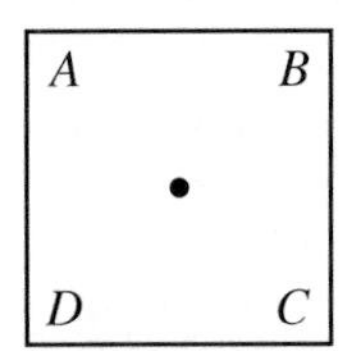

Figure 2.12: The square labelled differently.

In this last case the inner dancers swap places as do the two outer dancers. Now remember from earlier:

$$dc = aa \quad \text{so that} \quad dcdc = aaaa = e.$$

Here is the effect of repeating the operation dc four times.

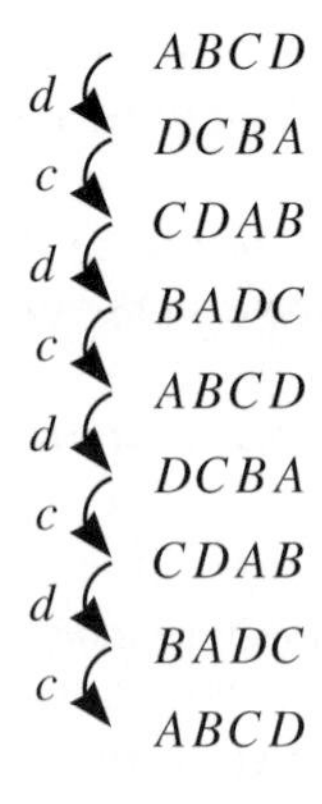

Figure 2.13: Coloured threads make this look like a knitting pattern.

If A and C are now stitches in red wool and B and D are stitches in blue wool and we continue this pattern for several more rows then we get the pattern seen in Figure 2.13, which you might recognise as part of a tartan. Here we see an application of mathematics to the design of the costumes in Scottish folk dancing. For an application to the costumes in English morris dancing, see the exercises.

Now use your imagination and a bit of group theory to design ever more exotic knitting patterns, dances, and bell-ringing patterns.

2.5 Exercises

First session

1. Symmetry can be found all around us. Think of as many possible examples as you can of objects which have symmetry (which can be symmetry under reflection, rotation, or translation).
2. Write down all the symmetries of the square and convince yourself that there are four rotations e, a, aa, aaa and four reflections, the three that we have already met, b, c, d and a new one f.
3. We have seen that $ab = d$, $ba = c$, $bc = a$, and $dc = aa$. See if you can work out how the other operations combine. For example, what are ac, ad, bd, etc? If you have time you can work out the 64 different ways of combining two possible operations. This is called the *group table*. Without necessarily doing this, can you work out some general rules, for example, what do you get if you combine two reflections, a reflection and a rotation, or two rotations?

4. Check that the rotations and reflections of a square really do form a group. To do this you need to complete the table in question 3 above and then look back and check the four rules for a group are really satisfied.
5. See if you can repeat questions 2 and 3 for either an equilateral triangle or (for the ambitious) a regular hexagon.
6. By bisecting each side, divide an equilateral triangle into six equal sections. Draw a random pattern in one of these sections. Now, using tracing paper if necessary reflect this pattern in each of the bisectors and continue to reflect what you have until you have drawn a pattern in the whole triangle. This will give you a pattern based upon symmetry that will always look good. This is the basic principle behind the kaleidoscope.

Second session

1. See if you can use the symmetries of the triangle to help you to design a dance for three people.
2. Notice how the two different ways of labelling the corners of the square in Figures 2.1 and 2.12 lead to different representations of the reflection d as a rearrangement of the sequence $ABCD$. What is the difference in rearrangement corresponding to the reflection b and the rotation a? Write out the sequence of moves corresponding to the reel $bcbcbcbc$.
3. Repeat question 2 for some other labellings of the corners of the square.
4. Create your own *knitting patterns*.

   We have generated several sequences of letters, both in the text and in the above question. Give each letter a different colour and use these to create some attractive 'knitting' patterns. For example, see what patterns you get for the simple operation of the repeated rotation $aaaaaaaa$ in the two cases of the corners arranged either in Figure 2.1 or Figure 2.12. (Hint: the second pattern could be called stripes.)

   A good way to do this is to use some sort of paper with a grid (such as graph paper) and to colour in the squares of the grid corresponding to the sequences of the letters.
5. In the American 'Grand square' dance, there are eight dancers who stand at the vertices of an octogon. If we label them clockwise round the octogon they are: $ABCDEFGH$. There are three basic moves a, b, c such that

$$a\colon ABCDEFGH \to HCBEDGFA, \quad b\colon ABCDEFGH \to EFCDABGH$$
$$c\colon ABCDEFGH \to ABGHEFCD.$$

   (a) Draw the octogon and see what the effect of each of these moves is. Are any a reflection or a rotation?
   (b) Show that $aa = bb = cc = e$.
   (c) A dance sequence is abc. How many times do the dancers have to repeat this sequence to get back to their original positions?

6. The dance 'Nottingham Swing' can be done with six dancers $ABCDEF$ (or indeed six couples, in which each letter represents a couple). There are two basic moves which are repeated to give the dance (together with some swings which do not change the order of the dancers). These moves are very similar to the inner and outer twiddle and are given by

$$ABCDEF \to BADCFE \quad \text{and} \quad ABCDEF \to ACBDEF.$$

   Find a way of labelling a *hexagon* so that these two moves correspond to reflections of the hexagon, and that the combination of the two moves corresponds to a rotation of the hexagon. (Hint: use a very similar pattern to the labelling of the square in Figure 2.1.)
7. Use the effect of rotations and reflections of the hexagon to devise some other dances for six dancers.
8. Go back to the section on bell-ringing. Write down all the possible bell-ringing moves on four bells $ABCD$. Do these moves form a group? If so write out a group table. If not explain why this set fails to be a group.
9. With a set of bells, try ringing a foursome reel.
10. This is a real problem in Morris dancing posed to us by a member of a Morris group. This group has twelve members and they want to design waistcoats for the group which are to be a 12×12 grid of squares made up of patches of twelve different colours. Can you design a waistcoat for them using these colours? The rules are that the waistcoats should have each colour once, and only once, in each row and column. The solution to this problem uses Latin squares and you can read about these in *Mathematical Recreations and Essays* by Rouse Ball and Coxeter.

Field trips and projects

Attending or organizing a folk dance, barn dance, square dance, or Ceilidh is not a traditional part of a mathematics course, but they are great fun and good exercise. Why not go wild and have an evening out? You can secretly record the moves of the dances and then analyse them when you get back. In the UK the English Folk Dance and Song Society (EFDSS) is a great source of information on local folk dance events. The Hobby Horse Club is a sub-branch of this for young people. There is also no reason why you can't try some of the exercises in this chapter to music to have some fun amongst yourselves.

2.6 Further problems

1. **The jumper group**
   Imagine that you have a woolly jumper. A moment's thought will convince you that there are four ways of wearing it: the usual way, inside out, back to front, and both inside out and back to front. Now imagine that you take your jumper off

every night, throw it into the corner of your bedroom and then put it back on again the next day.

We denote the normal jumper state by 'e', back to front by 'i', inside out by 'j' and both inside out and back to front by 'k'. Then, if we write the process of combining these as '$\star$' we can form a group! How your jumper ends up will depend on how it was to begin with and what happens when you threw it away the previous night. For example, if you started with it back to front 'i' and you throw it down so that it lands back to front 'i' when you put the jumper on again it will be normal 'e'. We can write this as $i \star i = e$.

This forms the following group:

$\star$	e	i	j	k
e	e	i	j	k
i	i	e	k	j
j	j			
k	k			

i. Look at the table where some of the entries have been filled in for you. Interpret what this means in terms of your jumper.
ii. Copy out and fill in the above table.
iii. Is this a *group*? The only way to satisfy a mathematician is to check the rules we gave earlier. Check all the rules.
iv. Now imagine you have a pair of socks and two feet. Each sock can be inside out or on the other foot, or both of course. This also forms a group. We could call it the *sock group*. Decide what all the combinations should be called and then write out a group table.
v. Can you extend this? (Imagine that you are a centipede!)

2. **The Clock Group**

This is an important example of a group. We will use this again later when we look at codes. Take a normal clock with hours that run from 1 round to 12 at the top. Imagine it is 9 o'clock. What time will it be in eight hours? The answer is, of course, 5 (not 17!). Remember, to be a mathematical group we need two things: a set and an operation on it.

- The set is the set of hours: $1, \ldots, 12$.
- The operation is adding time.

We will indicate this sort of adding by writing (mod 12) at the end of the sum to show that it is different. Thus $8 + 9 = 5 \pmod{12}$ but $8 + 9 = 17$.

i. Copy out and complete the following group table for the clock group.

	12	1	2	3	⋯	10	11
12							
1							
2							
3							
⋮							
10							
11							

ii. Why have we put 12 in the first column and row of the table?
iii. Check the other properties to make sure this is a genuine group.
iv. We can have other groups. Draw a clock with seven hours and write out a group table for it. Hint: $5 + 4 = 2 \pmod 7$. (Although you haven't seen a clock with seven hours, can you think of a connection with days of the week?)

2.7 Answers

First session

2. The new reflection f is through a diagonal leaving B and C of Figure 2.1 unchanged.
3. $ac = b, ad = f = ba^2$, and $bd = a^3$. In fact you only need to use a and b to write all the other symmetries and this is what we have done in the table below. The table shows xy. That is you read down first then along. Remember ba means *apply b to the square first then a*. Some books will write this the other way round so always check! A few seconds with a paper square will convince you that $ba = a^3b$.

$x\backslash y$	e	a	a^2	a^3	b	ab	a^2b	a^3b
e	e	a	a^2	a^3	b	ab	a^2b	a^3b
a	a	a^2	a^3	e	ab	a^2b	a^3b	b
a^2	a^2	a^3	e	a	a^2b	a^3b	b	ab
a^3	a^3	e	a	a^2	a^3b	b	ab	a^2b
b	b	a^3b	a^2b	ab	e	a^3	a^2	a
ab	ab	b	a^3b	a^2b	a	e	a^3	a^2
a^2b	a^2b	ab	a	a^3b	a^2	a	e	a^3
a^3b	a^3b	a^2b	ab	b	a^3	a^2	a	e

4. There are four rules we need to check to prove that the rotations and reflections of a square really do form a group. It is very important we do this before being satisfied that we are really dealing with a group and checking them almost always gives us a better idea of how the group works.

(i) The identity element of the group is e, as you can see from the group table.

(ii) It is clear from the table that if we combine any two elements of the group we get another one *in the group*. To check this you almost always need to write out the table.

(iii) Next we must find inverses. Take any element of the group and find it in the left-most column of the group table. For example a^3 is five rows down. Now look along the row and try and find the identity element you identified in 1 above. In our example this happens in the column a. So $a^3a = e$. We could also say that a^3-inverse is a or $(a^3)^{-1} = a$. Another example is ab; this has ab as an inverse. You can find the other inverses for yourself from the table but the important thing is that *every element has an inverse*.

(iv) Last we need to check, for any three elements x, y, and z, that $(xy)z = x(yz)$ (the brackets matter here). This is often the hardest to check but fortunately in mathematics you have to work very hard to find a set in which it fails! We need to satisfy ourselves that if we take three of our rotations and reflections this holds.

5. The reflections and rotations of the triangle and hexagon also form a group. They are called D_3 and D_6. A good way to write out the group tables is to call the smallest rotation a, a reflection of your choice b, and then to express all the other symmetries using these, just as we have for the square in the solution to problem 3 above.

Second session

2. Remember that in Figure 2.12 we are now reading the labels from the square *clockwise*. Thus the effects of the two symmetries are:

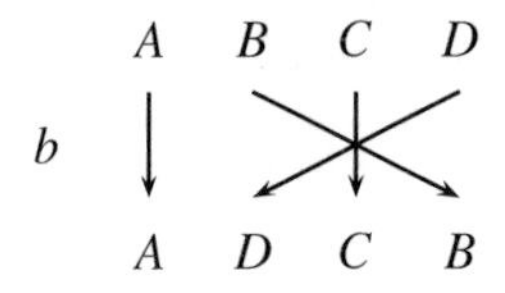

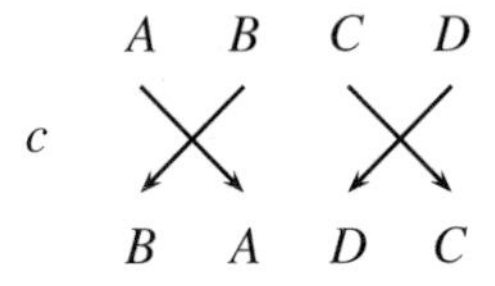

The effects of *bcbcbcbc* is:

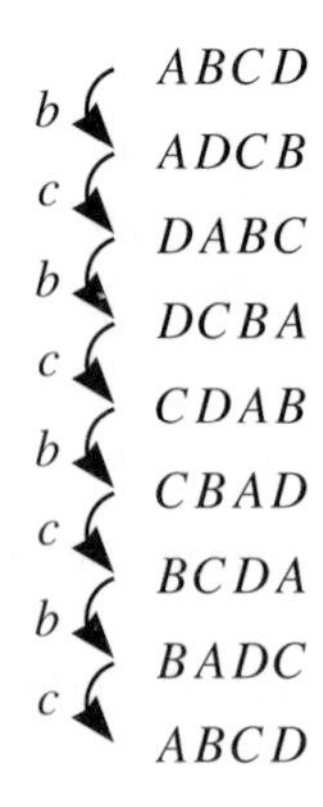

4. The knitting pattern for the sequence of moves worked out in question 2 is shown below:

5. In the American Grand square dance, the three basic moves can be thought of as follows:

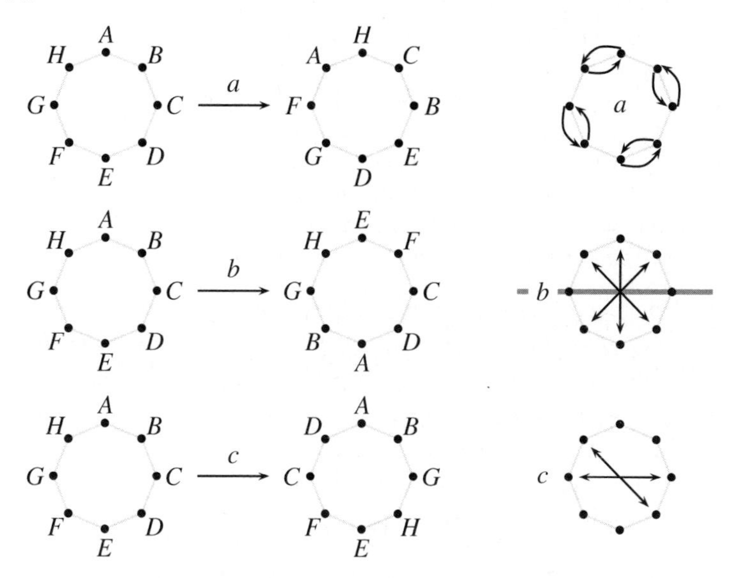

Only b is a reflection or rotation. From the pictures on the right it should be clear that $aa = bb = cc = e$. When we consider the dance sequence abc we have

$$\begin{aligned} ABCDEFGH &\rightarrow HCBEDGFA \\ &\rightarrow DEBCHAFG \\ &\rightarrow DEFGHABC \end{aligned}$$

Thus everyone has moved three steps anticlockwise. Eight repetitions of *abc* will be needed for everyone to get back to their original positions.

6. With the Nottingham Swing if we label the corners of a hexagon thus:

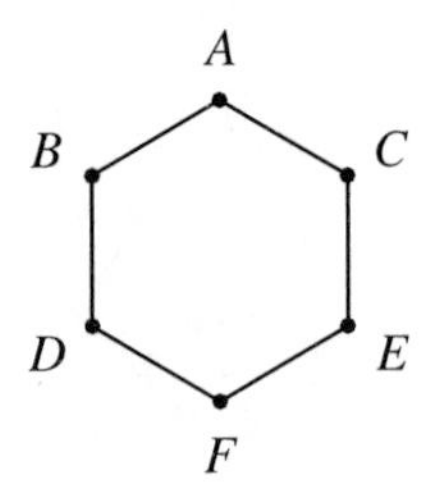

then the two operations are reflections, and the combination is a rotation, as shown:

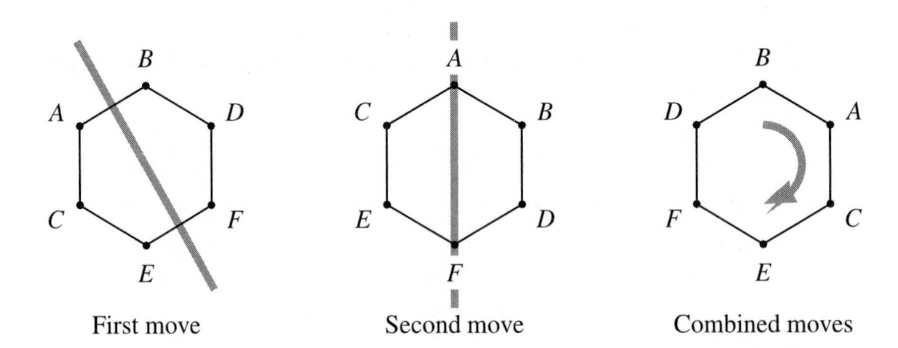

8. The bell-ringing moves are ones in which no bell moves more than one place. On $ABCD$ they are: $ABCD$ (which does nothing!), $ABDC$, $BACD$, $BADC$, and $ACBD$. These do not form a group. Take the following bell-ringing sequence:

$$ABCD \to BACD \to BCAD.$$

If the set was a group it would be *closed* and so the move

$$ABCD \to BCAD$$

would belong to it. But you can see that the A has moved two places so unfortunately the bell-ringing moves do not form a group.

10. One solution to this is given by the multiplication group modulo 13 described in Chapter 6.

Further problems

1. **The jumper group**

(b) The complete group table is

$\star$	e	i	j	k
e	e	i	j	k
i	i	e	k	j
j	j	k	e	i
k	k	j	i	e

(c) This is a group. The element e is the identity. Clearly combining two elements gives us something in the group. Every element is its own inverse, that is to say $x \star x = e$ for all x and a short check shows us that the last property also holds.

(d) The sock group has eight elements but instead of writing them as eight letters we will use a string of three 1's and 0's, e.g. 101. The three digits in order will represent the following things being true:

(i) socks on the wrong feet,

(ii) sock A inside out, and

(iii) sock B inside out.

Thus 001 will represent the situation where the socks are on the correct feet and sock B is on inside out.

To combine two elements of the group, say 001 and 101, we take the two strings of digits one at a time and 'add' then using the special group table

	0	1
0	0	1
1	1	0

Thus $001 \star 101 = 100$. This is not the same as binary arithmetic as there are no 'carries'.

(e) The answer to part (d) looks complicated and perhaps we could have used eight letters but it is now much easier to generalize this and build groups with longer strings of 1's and 0's.

A centipede with a hundred feet will have 100 socks. There are 100! (factorial) different ways of putting these sock on; each sock could be inside out. This group is rather large; in fact it is much larger than the number of particles in the known universe.

2. **The clock group**

(a) The complete group table for the clock group is:

	12	1	2	3	4	5	6	7	8	9	10	11
12	12	1	2	3	4	5	6	7	8	9	10	11
1	1	2	3	4	5	6	7	8	9	10	11	12
2	2	3	4	5	6	7	8	9	10	11	12	1
3	3	4	5	6	7	8	9	10	11	12	1	2
4	4	5	6	7	8	9	10	11	12	1	2	3
5	5	6	7	8	9	10	11	12	1	2	3	4
6	6	7	8	9	10	11	12	1	2	3	4	5
7	7	8	9	10	11	12	1	2	3	4	5	6
8	8	9	10	11	12	1	2	3	4	5	6	7
9	9	10	11	12	1	2	3	4	5	6	7	8
10	10	11	12	1	2	3	4	5	6	7	8	9
11	11	12	1	2	3	4	5	6	7	8	9	10

(b) We have put 12 first as it is the identity element. Unfortunately 0 doesn't appear on clocks; this would make the mathematics neater.

(c) The inverse of 12 is 12. The inverse of x is $12 - x$ for all the other elements.

(d) The group table for a seven-hour clock, also the same for adding days of the week, is:

	7	1	2	3	4	5	6
7	7	1	2	3	4	5	6
1	1	2	3	4	5	6	7
2	2	3	4	5	6	7	1
3	3	4	5	6	7	1	2
4	4	5	6	7	1	2	3
5	5	6	7	1	2	3	4
6	6	7	1	2	3	4	5

2.8 Mathematical notes

As mentioned in the text, the group formed by the square is better known as D_4, the *dihedral group* of four elements. There are other dihedral groups for all $n \geq 2$ formed by taking a polygon and making a group from the rotations and reflections.

It is always the case in D_n that $a^n = e$ and $ba = a^{n-1}b$ where a is the smallest rotation and b a reflection. This is useful to know when drawing the group table.

The clock group in the further problems is better known as C_{12} or the *cyclic group* with 12 elements. There are cyclic groups C_n of all orders n. Just think of a clock with a different number of hours. These are the simplest and most commonly encountered groups. Although we have not done so for the clock group it is better to use 0 and not n in the group. Thus C_7 would perhaps be better as the set of elements $\{0, 1, \ldots, 6\}$ rather than $\{1, \ldots, 7\}$. Of course it makes no real difference but is more a matter of taste.

The jumper group is also known as V_4 or the *Klein group*. This is an important group as it is the smallest non-cyclic group.

Another important example is the *symmetry group* S_n of n objects. This group can be formed from all the different ways n things can be arranged in a line. Thus S_n has $n!$ elements. You can think of D_4, the symmetries of a square, as being a small part of S_4 if you like. Just think of all the other ways of placing the letters on the corners of a square that you can't achieve with just rotations and reflections. This could lead us naturally onto the idea of one group being inside another but that, of course, is where the subject starts and we choose to finish!

2.9 References

Books

- Rouse Ball, W. W., and Coxeter, H. S. M. (1987). *Mathematical Recreations and Essays,* Chapter IX. Dover.
- Hamilton, J. (1978). *Know the Game: English Folk Dancing.* EP Publishing.
- Sharp, C. J. (1927). *The Country Dance Book: Part III: Containing Thirty Five Country Dances from The English Dancing Master (1650–1728).* Novello & Co, London.

  The above two books contain a comprehensive introduction to folk dancing.

  There are many good books on groups. Most are aimed at university-level mathematicians but don't let that put you off.
- Jordan, C. R., and Jordan, D. A. (1994). *Groups.* Edward Arnold.
- Fraleigh, J. B. (1994). *A First Course in Abstract Algebra.* Addison Wesley.
- Hall, F. M. (1980). *An Introduction to Abstract Algebra,* Volume 1. Cambridge University Press.

Web sites

- `http://www-groups.dcs.st-and.ac.uk` gives a section on the history of group theory.
- `http://www.math.niu.edu/~rusin/known-math/` Look in the Mathematical Atlas.

3 Sundials: how to tell the time without a digital watch

3.1 Introduction

OK, so you have just got the latest in digital watches, with a stop watch, date, times from all over the world, and the ability to function at 4000 fathoms. Just one small problem, the batteries have gone flat. If the sun is shining, however, you don't need to use a watch at all, because the sun makes an excellent clock which (fortunately) doesn't need batteries that can run down. This masterclass is all about how we use sundials to tell the time, and contains real-life construction projects, including a splendid garden ornament and the perfect way to decorate the school playground. With some care, and using some interesting mathematics, we can end up building a very accurate sundial. If you find the mathematics in the first sections heavy going then rush on to the later sections where you will see how to construct a variety of different types of sundial. However, we hope that you will enjoy the mathematics as well.

3.2 A brief history of telling the time

Although we often take the problems of telling the time or of finding the day's date for granted, finding solutions to these are vital to our civilization. Early men and women needed to have an accurate notion of the seasons in order to know when to plant and harvest their crops. As civilization developed it became important to know the time of the day. Two times were obvious, as everyone was aware of when the sun rose (dawn) and when it set (dusk). More accurate observations of the sun then showed that having risen in the east it climbed to its highest point in the sky, which is due south (in the northern hemisphere), before descending towards its setting point in the west. The time when the sun is at its highest point (which is when the shadows are shortest) is called noon. Knowing dawn, noon and dusk separated the day into morning, afternoon, and night. The Chinese were probably the first people to define morning and afternoon and they gave them the rather splendid names of *before horse* for morning, *horse* for noon, *dog* for evening. Midnight then became *rat*. In China early time measurements were sometimes made by burning sawdust in a pit in the ground and following the progress of the fire throughout the day. This was not a particularly accurate way to tell the time and, of course, did not work in the rain. Later on, around 1500 BC the Chinese started to use the sun as a more accurate means of telling the time and were the first people to record the use of the sundial. Sundials also appear to have been mentioned in the Bible. In 2 Kings, Chapter 20, verses 9–11, what sounds rather like a sundial is described. Here it says

> *Isaiah prayed to the Lord, and the Lord made the shadow go back ten steps on the stairway set up by King Ahaz.*

It is thought (with some archaeological evidence) that the 'stairway' is a sundial and the 'steps' are the divisions. In these verses the shadow cast by the sun is going *backwards*, which is impossible in general and hence is an extraordinary and miraculous event, which must have greatly impressed King Hezekiah, who was watching. This stairway is called the 'dial of Ahaz' after the king of Judah who constructed it, and is thought to have been built around 700 BC. From then on sundials spread rapidly into the West and the Roman Vitruvius (a contemporary of Julius Caesar who we will meet in the chapter on codes) made what is possibly the first list of different types of sundial.

Around 2000 BC, as we will see in Chapter 7, the Babylonians made a series of careful observations of the sun and deduced that one year lasted approximately 365.25 days. Having derived this they could then construct the first calendar. This is a pivotal invention in the history of the world, as from the calendar they could tell what was the best time to plant certain crops. In fact calendars have a very interesting history in their own right and it took several refinements of the calendar before the present day system was settled upon. The Babylonians counted in base 60 and the number of days in the year is very close to six sixties or 360. By dividing the circle into 360 degrees they then had a measurement of angle in which each degree roughly corresponded to one day, and which was perfect from their standpoint of a number base.

In Europe during the Dark Ages, the measurement of time went into a decline although various attempts to record it were used, for example the burning of a coloured wax candle was used by King Alfred. In each candle there was a series of coloured rings, each ring taking about one hour to burn down. In the Middle Ages the first mechanical clocks were introduced into Europe. These were mostly installed in churches, as a church was a central building everyone could see, and also a church needed to tell people when to come to mass. The early clocks in the churches did not have hands, but instead rang a bell when it was time for the services. A splendid example of a clock (the original design of which dates from AD 1386) can be found in Salisbury Cathedral, Salisbury, England. This is well worth a visit (though you may raise a few eyebrows when you spend all of your time watching the clock rather than admiring the stained glass windows, etc.). The Salisbury clock is very large and uses heavy weights. It is also not very accurate.

More accurate and smaller clocks became possible following the discovery of the pendulum by Galileo in 1637. Galileo is one of the more colourful mathematicians of history who made many remarkable mathematical and scientific discoveries (such as the basic principles of mechanics, the moons of Jupiter, and the fact that a projectile travels in a parabola), as well as running foul of the Spanish Inquisition. During a church service he was bored, and seeing a swinging chandelier did what any self-respecting mathematician would, and timed the period of its swings using his pulse. He made the remarkable observation that almost regardless of the size of the swing the time of the swing was (according to his pulse) constant. (It is tempting to think that in the excitement of this scientific discovery, his pulse quickened thus possibly wrecking the observation!) The constancy of the time it took for a pendulum to make a small swing made it possible to use the pendulum in a clock, giving us the basic design of the grandfather clock.

A smaller device which kept time accurately and could be used in a portable watch was the spring escapement invented by the English mathematician Hooke. Clocks based on this idea were used almost until the present day and were brought to a point of perfection by John Harrison, who developed such an accurate watch (called the chronometer) that ships could use it to find their longitude at sea. Spring escapement watches have now been largely superseded by digital watches, which rely for their accuracy on the constancy of the electrical oscillations of a quartz crystal. These keep an accuracy of about one second a year – but are still no good if their batteries run out!

Although there are many ways to tell the time (see the exercises) the sundial probably holds the record for having been used for the longest, and it can still be as accurate as a good pocket watch if made and used correctly.

3.3 The motion of the sun

Shadows at noon

To tell the time by using the sun it is convenient to forget the discovery by Copernicus that the Earth goes round the sun, and to take the rather simpler view (for us) that the sun goes round the Earth. In this Earth-centred view, the sun rises in the east and sets in the west and traces a circular path in the sky during the day. If we record noon on two successive days, then the time interval between two noon points is called one day and this is divided up into 24 to give a definition for the hour. (In fact the length of time, measured with a watch, between two noons varies slightly during the year because the Earth travels on an elliptical orbit round the sun and its axis is tilted with respect to this orbit. We have to average the length of the day out over a year to give a definition for an hour. This leads to a difference in time measurement based on the sun which we will come back to in Section 3.7.)

Until we reach Section 3.7 we will assume that one day lasts exactly 24 hours as measured on a clock. It is inconvenient (and indeed dangerous) to measure the position of the sun by looking at it directly, and a much better way to record its position is by looking at the shadow that it casts.

Remember:
Never look directly at the sun!

Suppose that we set up a vertical pole on a horizontal surface on a sunny day. A person standing up makes a very good pole. The pole (person) will cast a shadow, and if you watch over a period of a few hours (if the person can stand in one place for long enough), the shadow will move round. (The shadow can easily be recorded by placing a marker at different times of the day.) In fact the shadow moves in a *clockwise* direction in the northern hemisphere. This is *exactly* why the first mechanical clocks have hands that rotate in this way and it is from this we get the term 'clockwise'.

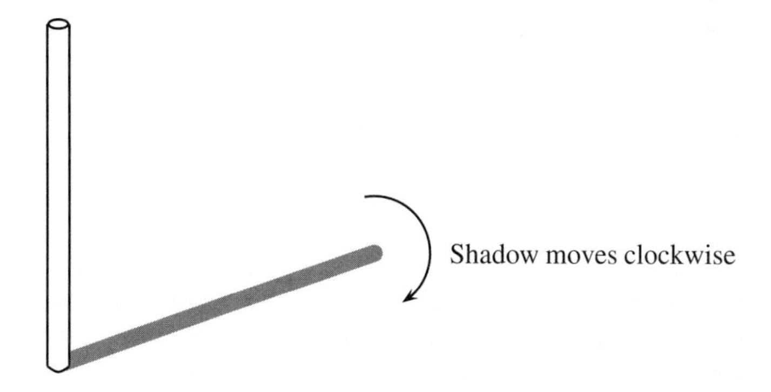

The length of the shadow is shortest precisely at noon when it points directly north (or in the southern hemisphere, south). One way to find the noon point is to measure the shadow at some point in the morning – drawing a line OA as indicated in Figure 3.1 (where O marks the position of the base of the pole). Now wait till the afternoon and draw another line OB when the shadow is exactly the same length as it was in the morning. You can tell by drawing a circle centred at O, of radius OA, and waiting until the shadow from the end of the stick crosses it. If you bisect the angle between the two lines OA and OB then you will have a line OC which marks the position of noon. You can mark this line as a way of always knowing when it is noon. These lines, called 'noon marks' or 'meridian marks', are quite common on older buildings, and could be used to set your watch with (although there is always the difference between true noon and 12 o'clock as measured on a watch, which we will come to in Section 3.8).

Even if the pole always has the same length, the length of the noon shadow OC varies during the year. It is longest on mid-winter's day on December 22nd and

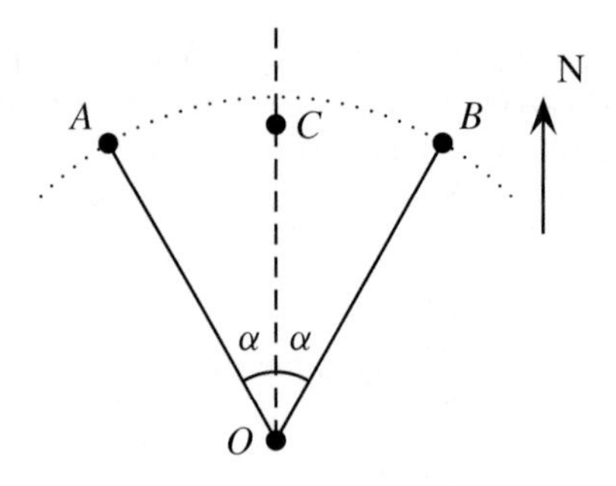

Figure 3.1: Finding true north.

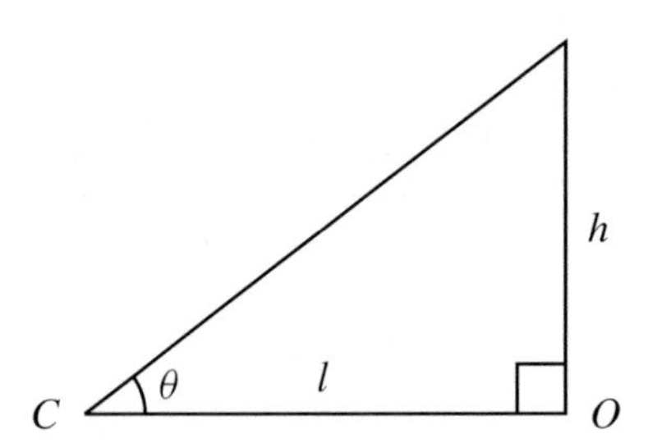

Figure 3.2: Finding the angle of the sun from the horizon.

shortest on mid-summer's day on June 21st. The reason for this is that the angle that the sun makes with the horizontal changes during the year. Indeed, it is this change in angle that causes the difference in the seasons. To find this angle, measure the length l of the shadow OC and the height h of the pole. The angle (which we will refer to as θ, pronounced *theta*) is shown in Figure 3.2.

Using your calculator you can work out the angle using the *inverse tan* function given by the $\tan^{-1}$ button (on some calculators, and in tables, this is labelled *arctan*) by using the formula

$$\theta = \tan^{-1}(h/l).$$

For example, if the height of the pole is 50 cm and the length of the shadow is 1 m then the angle the sun makes with the horizontal is

$$\theta = \tan^{-1}(1/2) = 26.5^{\circ}.$$

The value of θ not only depends upon the time of year but also upon your *latitude*. We will use the Greek letter ϕ (pronounced *phi*) to denote your latitude throughout this chapter. It is important that you find out the latitude for your own location, otherwise you cannot construct an accurate sundial. You can find your latitude (together with your longitude – which you will need later) from an atlas. In Bath the latitude is

Table 3.1: Locations of major world cities outside the United Kingdom.

City	Latitude	Longitude
Beijing	39.9 N	116.4 E
Bombay	18.9 N	72.9 E
Berlin	52.5 N	13.5 E
Boston	42.3 N	71.1 W
Buenos Aires	34.7 S	58.4 W
Cairo	30.4 N	31.2 E
Calcutta	22.6 N	88.3 E
Cape Town	33.9 S	18.5 E
Delhi	28.7 N	77.2 E
Dublin	53.4 N	6.3 W
Jakarta	6.1 S	106.8 E
Los Angeles	34.0 N	118.2 W
Melbourne	37.7 S	145.0 E
Mexico City	19.3 N	99.2 W
Moscow	55.8 N	37.7 E
New York	40.7 N	35.8 W
Paris	48.8 N	2.2 E
Rome	41.9 N	12.5 E
São Paulo	23.6 S	46.7 W
San Francisco	37.7 N	122.7 W
Seattle	47.6 N	122.2 W
Shanghai	31.1 N	121.5 E
Singapore	1.2 N	103.7 E
Seoul	37.4 N	127.0 E
Sydney	33.9 S	151.2 E
Tel Aviv	31.1 N	34.8 E
Tokyo	35.6 N	139.7 E
Vancouver	49.1 N	123.1 W
Washington DC	38.9 N	77.0 W

51.5°N and the longitude is 2.5°W. For convenience, the latitude and longitude of some major cities around the world are shown in Table 3.1 and those of some cities in the UK are shown in Table 3.2, with angles expressed in degrees and fractions of degrees.

Knowing ϕ, the precise value of θ (and indeed the angle α in Figure 3.1) can be found at any time of the year using a bit of spherical trigonometry. This is hard stuff and the formulae that you obtain are rather messy.

However at mid-summer and mid-winter it is rather simpler to work out θ and we will see how to do it in the further problems. What we find from this calculation is that if θ_{summer} is the angle of the sun on mid-summer's day and θ_{winter} the angle on

Table 3.2: Locations of major cities in the United Kingdom.

City	Latitude	Longitude
Aberdeen	57.1 N	2.1 W
Belfast	54.6 N	6.0 W
Birmingham	52.5 N	1.9 W
Bristol	51.4 N	2.6 W
Cambridge	52.2 N	0.1 E
Cardiff	51.5 N	3.2 W
Derby	52.8 N	1.4 W
Dundee	56.5 N	3.0 W
Edinburgh	56.0 N	3.1 W
Glasgow	55.8 N	4.2 W
Hull	53.8 N	0.2 W
Leeds	53.8 N	1.5 W
Leicester	52.7 N	1.2 W
Liverpool	53.4 N	3.0 W
London	51.5 N	0.0
Manchester	53.5 N	2.2 W
Middlesborough	54.6 N	1.2 W
Newcastle upon Tyne	55.0 N	1.5 W
Nottingham	53.0 N	1.2 W
Oxford	51.8 N	1.2 W
Plymouth	50.4 N	4.2 W
Sheffield	53.4 N	1.4 W
Southampton	50.8 N	1.3 W
Stoke-on-Trent	53.0 N	2.1 W
Swansea	51.5 N	4.0 W

mid-winter's day then

$$\theta_{\text{summer}} = 113.5^\circ - \phi \quad \text{and} \quad \theta_{\text{winter}} = 66.5^\circ - \phi.$$

The angles 113.5° and 66.5°, which look rather mysterious, are derived from the fact that the axis about which the Earth rotates is not perpendicular to the plane of its orbit round the sun, but is tilted at the angle of 66.5^o to the plane of its orbit around the sun. In summer the northern hemisphere points towards the sun, and in winter it points away from the sun. We then have

$$113.5^\circ = 180^\circ - 66.5^\circ.$$

The *average* value θ_{av} of θ over the year is the mean of the mid-summer and mid-winter angles and is given by

$$\theta_{\text{av}} = 90^\circ - \phi.$$

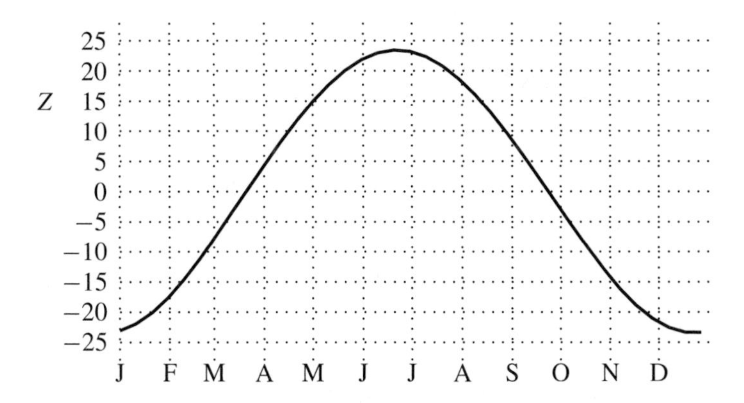

Figure 3.3: The declination of the sun.

During the year the value of the angle θ at any point in the northern hemisphere is given by

$$\theta = \theta_{\text{av}} + Z.$$

The angle Z is called the *declination* of the sun, with $Z = 23.5^\circ$ on mid-summer's day and $Z = -23.5^\circ$ on mid-winters day. Figure 3.3 is a graph of Z throughout the year.

This graph oscillates either side of a mean value of zero. There are two dates in the year when it is at this angle. These two dates are called the spring and autumn *equinoxes* and occur on March 21st and September 23rd respectively (with slight variations during a leap year). The equinoxes mark the transition from spring to summer and from summer to autumn. These dates are also roughly when the clocks change from winter time to summer time and vice versa.

We will need to know Z later when we come to lay out an analemmatic sundial. Its value in degrees at different times during the year is shown in Table 3.4 in Section 3.6.

The rest of the day (and night)

We have now worked out what the sun does at noon. What does it do at other points in the day? To work this out we must consider a point in the sky directly above the North Pole. This is the point which the axis of the Earth points to and in the northern hemisphere is conveniently marked by the Pole Star – *Polaris*. (In fact Polaris is not quite at the North Pole, but it is close enough for our purposes.) Unfortunately, there is no such good marker in the southern hemisphere.

To find Polaris, wait till a clear night. In the northern part of the sky there is a bright constellation called variously the Great Bear, the Plough, and the Big Dipper. This resembles a large bear with a body of four stars and a tail of three stars. Take the two bright stars on the end of the body furthest from the tail (called the pointers) and draw

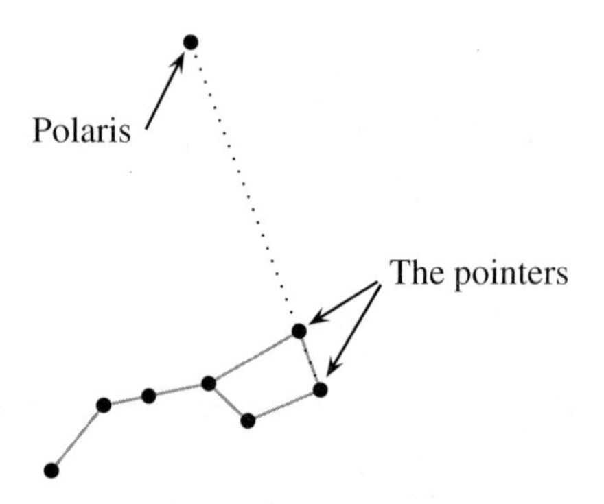

Figure 3.4: Finding Polaris using the stars.

a line through them in the sky. If you follow this line up then the first bright star that you come to is Polaris. (In fact Polaris is at the end of the tail of another constellation called the 'Little Bear', but this is a rather dimmer constellation and not easy to spot.)

Exercise. *Polaris is the only star in the sky which does not appear to move during the night or during the year. Can you work out why?*

This important fact was known to the ancients who made careful observations. Finding Polaris was essential to early navigators as once found they could accurately locate north and then steer in the right direction. There is good evidence that some species of birds use Polaris to navigate with. As well as pointing to north, Polaris also helps you to work out your latitude.

> Your latitude is the angle that Polaris makes with the horizon.

This fact is illustrated in Figure 3.5 and it is not too hard to work out why it is true. The mathematics behind it is covered in the first of the further problems. If you are interested in finding out more about the fascinating history of navigation then read the lovely book *Longitude* by D. Sobel.

The main interest that we have in Polaris is that it marks the point in the sky around which the sun appears to rotate. In fact, to a very good approximation, in 24 hours the sun appears to move all the way around a large circle in the sky which is centred on Polaris. Of course during the day we can't see Polaris because the light from the sun drowns it out! The diameter of this circle changes during the year. If we could see the stars during the day, then we would see the sun change its position relative to the fixed stars from one day to the next. It moves on a line called the ecliptic which

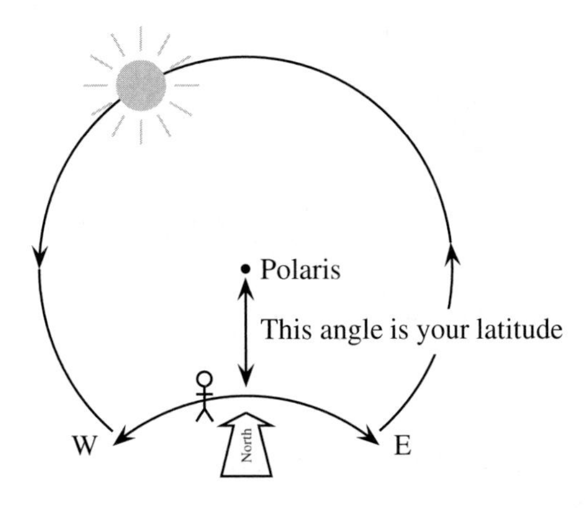

Figure 3.5: Your view looking north.

passes through the constellations of the zodiac. The date in the year is marked by which constellation the sun is currently in. The sky is bisected by a circle at an angle of 90° to Polaris. This is the Celestial Equator, and it is aligned to be parallel with the Earth's Equator. During summer the sun is *above* the Equator (i.e., it is between the Equator and Polaris). At the spring and autumn equinoxes it is *on* the Equator and in the winter it is *below* the Equator. In fact the angle between the sun and the Equator is the declination Z which we learnt about earlier. A picture of the celestial sphere on which the sun appears to move is given in Figure 3.6.

Suppose that we were to place a pointer in the ground at such an angle and direction that it pointed directly toward Polaris. Now imagine that a disk of card is stuck to this pointer so that it is at right angles to it. This piece of card is now aligned with the Equator. Because the sun goes round the sky in a circle centred upon Polaris, it casts a shadow on the card which *has the same length at every time during the day* and looks exactly like the hour hand of a clock, moving clockwise around the card. An hour hand on a clock moves twice around the clock in 24 hours, but the shadow cast by the sun moves at half this speed and makes a complete revolution of 360° in 24 hours. Thus it moves at a constant speed of $360°/24 = 15°$ every hour.

> The shadow cast by the sun looks like an hour hand moving at a rate of 15° per hour.

By using such a pointer and a disk aligned with the Equator we have the basis of our first sundial – the *Equatorial sundial.*

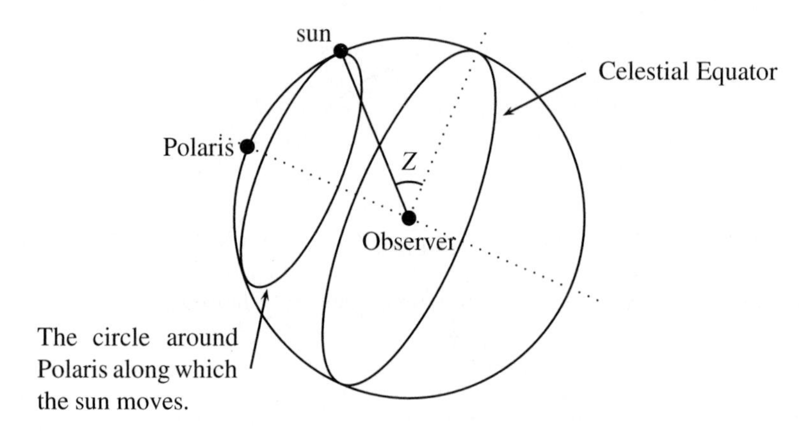

Figure 3.6: The celestial sphere.

3.4 The Equatorial sundial

Before you start making your sundial take a quick look at the completed Equatorial dial in Figure 3.8. To start the construction draw a circle on a stiff sheet of card and cut round this to make a disk. In principle this can be of any diameter, with about 10 cm being a good size for a simple sundial. Mark the disk out in equal divisions of 15°: this will give you 24 divisions in all. Number these divisions starting from 1 and working around *clockwise* till you get to 12. You can then either continue to 13 (for a 24-hour clock) or start again from 1. In the latter case you will get a disk like the one in Figure 3.7(a).

Figure 3.7(a): Front

Figure 3.7(b): Back

Now, turn over the disk and draw divisions on the other side, making sure that the divisions on the back line up precisely with the divisions on the front. Mark the back division at 12 in exactly the same place as the front division also marked 12, and mark the rest of the divisions as shown in Figure 3.7(b), such as 1, 2, etc., but go round *anticlockwise* starting from 12. Next make a small hole in the centre. Take a knitting needle or some other sharp pointed object such as a kebab stick and *carefully* push it through the hole from the back of the disk to the front. For the sundial to work, the length of this stick below the dial is important. If the radius of the disk is r and the length of the stick below the disk is l then l is given by the formula

$$l = r/\tan(\phi).$$

The hole needs to be small enough to keep the disk secure on the knitting needle. If the disk wobbles secure it with some glue or sticky tape. It is most important that the needle should be perpendicular to the disk and you should check this with a set-square.

We now have to mount the Equatorial sundial on a base. Firstly stick the bottom of the knitting needle to a piece of strong card on a horizontal surface using a dab of Blu-tack. Having done this rotate the disk until the division marked 12 on both the top and the bottom is the point where the disk touches the cardboard. This point will be where the shadow of the sun will fall at noon. You can now stick this point to the card with some more Blu-tack.

Exercise. *Show that choosing* $l = r/\tan(\phi)$ *places the pointer at an angle of* ϕ *to the horizontal.*

To set up your sundial, take it to a suitably sunny place. Make sure the base is horizontal and rotate the base card till the needle points due *north*. You can do this during the day by using a compass or during the night by pointing the needle towards Polaris. The disk is now aligned with the Equator, and this gives the sundial its name. Hopefully Figure 3.8 makes all this clear!

Your Equatorial sundial will now tell the time.

When the sun is shining, the pointer of the Equatorial sundial casts a shadow onto the disk. The division that the shadow falls on is then the time, with noon being when the shadow hits the 12 point at the base of the disk. During the summer the shadow will go around the top of the disk moving *clockwise* just like the hour hand of a watch.

Unlike the hour hand of a clock, which always has a constant length, the length of the shadow varies during the year. On mid-summer day it is at its shortest. It then lengthens, reaching its maximum length at the autumnal equinox. Remember that this is when the sun is on the Equator. Something interesting now happens as we move on past the autumnal equinox. The sun moves below the equator and the shadow it casts is now on the *bottom* of the disk. (This is why we had to number the disk on both

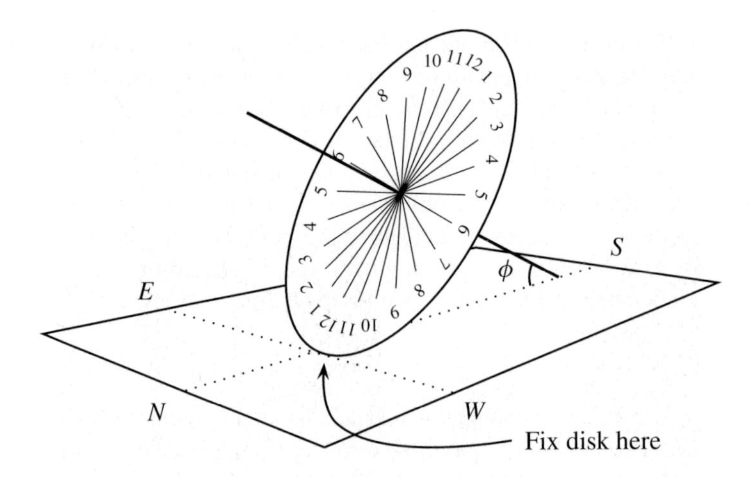

Figure 3.8: The completed Equatorial sundial.

sides.) At this point the Equatorial sundial becomes tricky to use and can give you a crick in the neck. The shadow of the pointer moves around the bottom of the disk anticlockwise and is at its shortest on mid-winter day (when you may not see the sun at all due to inclement weather or just staying in to watch the TV). It stays on the bottom of the disk until the spring equinox when it moves back to the top of the disk again.

The Equatorial sundial has a certain mathematical simplicity, has an easy to construct dial (just like a watch face) which is the same anywhere in the northern hemisphere, and the shadow cast by the sun has equal length during the day. These are desirable properties. However the Equatorial sundial also has big disadvantages. The dial can only be read conveniently for half of the year, and can only be at one angle. Thus we cannot mount the Equatorial sundial on a horizontal or a vertical surface – which is what we are most likely to find in practice. These disadvantages are overcome by the next sundial – the horizontal sundial.

3.5 The horizontal sundial

When you think about what a sundial looks like, then the chances are that you are considering a *horizontal* sundial. In such a sundial the base with the numbers on is horizontal (hence the name) and the pointer is a spike at an angle to the base that equals your latitude. For this section we will follow the convention for a horizontal sundial and refer to the pointer as a *gnomon*. Usually the gnomon is supported by a triangular construction decorated with cherubs or something similar.

The gnomon of a horizontal sundial is aligned along a north–south line. This line will be the mark for 12 noon, and you should mark 12 (or XII) at a convenient point due north of the base of the gnomon.

Figure 3.9: A multiple sundial designed and made by David Brown.

Constructing the base of a horizontal sundial is rather more complicated than constructing the base of an equatorial sundial, because the base is not perpendicular to the gnomon. This distorts the lines along which the shadow of the gnomon will fall so they are not at equally spaced angles around the dial. The shadow will also have different lengths for different times of the day. This is the price that we pay for the (relative) convenience of having a horizontal base.

Fortunately, the application of some mathematics will allow us to project the hour lines of the Equatorial sundial onto the hour lines of the horizontal sundial. By doing this we can draw a series of lines on the horizontal base so that the shadow cast by the gnomon will exactly pass through each line on its corresponding hour. Having done this the sundial can be used throughout the year, and the shadow will always fall on the top of the base (even during the winter months), provided of course that the sun is shining. A typical example of the base of a horizontal dial is given in Figure 3.10.

It comprises a series of lines labelled XII, I, II, III, IV, V, VI, etc., which are centred upon the point where the gnomon meets the base. It is traditional to use Roman numerals for a horizontal sundial and we have done this here. Each of these lines makes an angle x with the north–south line. This angle is 0° for the XII line and 90° for the VI line. However, unlike the Equatorial sundial, these angles are not simple multiples of 15° and to calculate the angle involves using a satisfyingly complicated formula that amply justifies the employment of a mathematician.

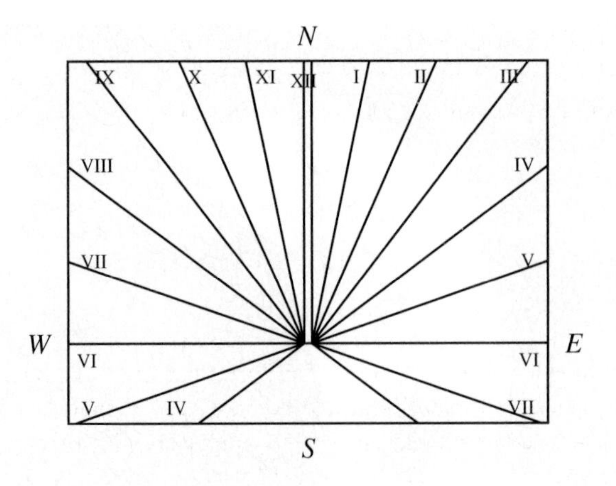

Figure 3.10: A horizontal dial calibrated for 52°N.

Interesting fact.

The dial that you construct depends upon your latitude, so that a horizontal sundial constructed for Bath will be different from one constructed for Edinburgh or for San Fransisco.

There are three ways in which you can construct the base of a horizontal sundial.

1. Copy an existing design. If you want to make a horizontal sundial quickly then this is the easiest way to do it, although it is lacking in mathematical sophistication. In Figure 3.22 we give an example of a horizontal design for a latitude of 52°N. To make a sundial, photocopy this design and then cut it out. If you cut along the dotted line, fold where indicated, and stick the paper tabs to a suitable base, then you have an instant horizontal sundial. The main problem with this is that the design will only work if your latitude is close to 52°N. At other latitudes it will be rather inaccurate.

2. Use a formula. You can construct a horizontal dial which will work in a general latitude by using a formula which employs the buttons that you will find on your pocket calculator. To construct the sundial using a formula, we will draw a series of hour lines which are projections of the hour lines on the Equatorial disk. This requires the use of some trigonometry and we show how it is done in the second worked example in the further problems.

Suppose that your latitude is ϕ and the time that you want to draw the line for is T hours. We will call $T = 0$ if we are at noon. In the afternoon T is just the time expressed in hours and fractions of hours, so that at 3.30 p.m. we have $T = 3.5$.

In the morning we find T by subtracting 12 from the hour so that at 10.00 a.m. $T = -2$. Now suppose that the noon line is at $0°$ and we want the hour line corresponding to T to be at x degrees to the noon line (measuring clockwise). Then the angle x that you need to draw is given by

$$x = \tan^{-1}[\tan(15T)\sin(\phi)]$$

where all angles are measured in degrees. This formula may at first look rather complicated, but using a standard scientific calculator you can calculate x using the buttons on your calculator marked sin, tan, and $\tan^{-1}$. For example, suppose that we want to work out the angle for an hour line at 4.00 p.m., so that $T = 4$, at a latitude of $\phi = 52°$. We proceed as follows:

- Multiply T by $15°$ to give $60°$.
- Find $\tan(60°)$. This is 1.7320.
- Find $\sin(\phi)$. For example if $\phi = 52°$ then $\sin(52°) = 0.7880$.
- Multiply $\tan(15T)$ by $\sin(\phi)$ to give 1.3649.
- Find $\tan^{-1}(1.3649)$. This is $53.7709°$.
- We have $x = 53.7709°$ which we can round to $x = 53.8°$. Mark the hour line for IV at this angle, measured clockwise from the line marking noon.

Exercise. *As a second example, at 9 a.m. we take $T = -45°$. Show that for $\phi = 52°$ the formula gives $x = -38.2°$.*

We plot the hour line for IX by measuring $38.2°$ anticlockwise from the noon line. We show this in the following figure

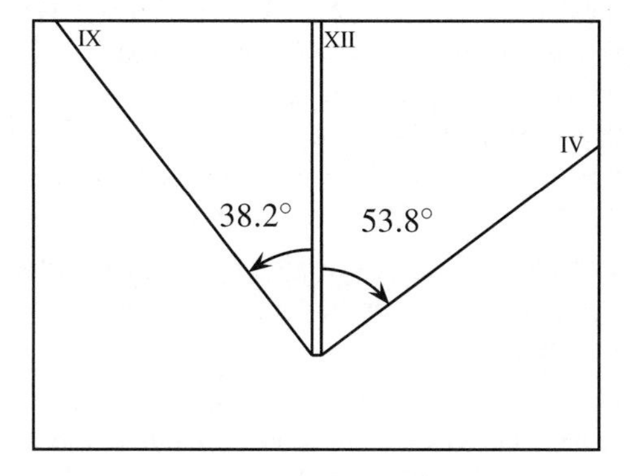

If you have access to a computer or a scientific calculator which you can program then it is not too hard to program this formula so that if you give T the program will return x. Do this for a selection of times during the day.

Table 3.3: Angles of hour lines for vertical and horizontal sundials in the northern hemisphere.

Latitude for a horizontal sundial

Time	10°	20°	30°	40°	50°	60°	70°	80°	90°
5.00 a.m.	−147.1	−128.1	−118.2	−112.6	−109.3	−107.2	−105.9	−105.2	−105.0
5.30 a.m.	−127.2	−111.1	−104.8	−101.6	−99.8	−98.6	−98.0	−97.6	−97.5
6.00 a.m.	−90.0	−90.0	−90.0	−90.0	−90.0	−90.0	−90.0	−90.0	−90.0
6.30 a.m.	−52.8	−68.9	−75.2	−78.4	−80.2	−81.4	−82.0	−82.4	−82.5
7.00 a.m.	−32.9	−51.9	−61.8	−67.4	−70.7	−72.8	−74.1	−74.8	−75.0
7.30 a.m.	−22.7	−39.5	−50.4	−57.2	−61.6	−64.4	−66.2	−67.2	−67.5
8.00 a.m.	−16.7	−30.6	−40.9	−48.1	−53.0	−56.3	−58.4	−59.6	−60.0
8.30 a.m.	−12.8	−24.0	−33.1	−40.0	−45.0	−48.5	−50.8	−52.1	−52.5
9.00 a.m.	−9.9	−18.9	−26.6	−32.7	−37.5	−40.9	−43.2	−44.6	−45.0
9.30 a.m.	−7.6	−14.7	−21.0	−26.3	−30.4	−33.6	−35.8	−37.1	−37.5
10.00 a.m.	−5.7	−11.2	−16.1	−20.4	−23.9	−26.6	−28.5	−29.6	−30.0
10.30 a.m.	−4.1	−8.1	−11.7	−14.9	−17.6	−19.7	−21.3	−22.2	−22.5
11.00 a.m.	−2.7	−5.2	−7.6	−9.8	−11.6	−13.1	−14.1	−14.8	−15.0
11.30 a.m.	−1.3	−2.6	−3.8	−4.8	−5.8	−6.5	−7.1	−7.4	−7.5
12.00 a.m.	0.0	0.0	0.0	0.0	0.0	0.0	0.0	0.0	0.0
12.30 p.m.	1.3	2.6	3.8	4.8	5.8	6.5	7.1	7.4	7.5
1.00 p.m.	2.7	5.2	7.6	9.8	11.6	13.1	14.1	14.8	15.0
1.30 p.m.	4.1	8.1	11.7	14.9	17.6	19.7	21.3	22.2	22.5
2.00 p.m.	5.7	11.2	16.1	20.4	23.9	26.6	28.5	29.6	30.0
2.30 p.m.	7.6	14.7	21.0	26.3	30.4	33.6	35.8	37.1	37.5
3.00 p.m.	9.9	18.9	26.6	32.7	37.5	40.9	43.2	44.6	45.0
3.30 p.m.	12.8	24.0	33.1	40.0	45.0	48.5	50.8	52.1	52.5
4.00 p.m.	16.7	30.6	40.9	48.1	53.0	56.3	58.4	59.6	60.0
4.30 p.m.	22.7	39.5	50.4	57.2	61.6	64.4	66.2	67.2	67.5
5.00 p.m.	32.9	51.9	61.8	67.4	70.7	72.8	74.1	74.8	75.0
5.30 p.m.	52.8	68.9	75.2	78.4	80.2	81.4	82.0	82.4	82.5
6.00 p.m.	90.0	90.0	90.0	90.0	90.0	90.0	90.0	90.0	90.0
6.30 p.m.	127.2	111.1	104.8	101.6	99.8	98.6	98.0	97.6	97.5
7.00 p.m.	147.1	128.1	118.2	112.6	109.3	107.2	105.9	105.2	105.0
	80°	70°	60°	50°	40°	30°	20°	10°	0°

Latitude for a vertical sundial

Note: You may have problems if the time is after 6.00 p.m. or before 6.00 a.m. as then $\tan(15T)$ is negative and the calculator may give a silly answer. You should subtract 180° to get the correct angle.

3. Use tables. If you are as old as we are (don't ask) then you will be familiar with using tables (such as log and sin tables). Although tables are a bit out of fashion now, they lead to a good way of constructing a sundial. What the tables do is evaluate the angle x by using the formula for a set of different latitudes and times. A suitable table is given in Table 3.3.

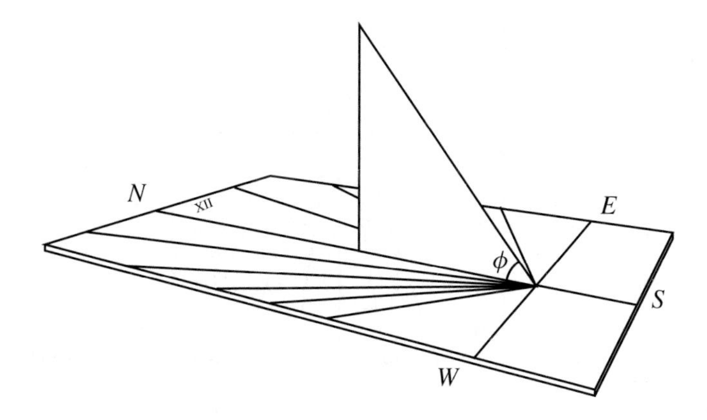

Figure 3.11: The completed horizontal sundial.

To use this table you need to know your latitude ϕ. Now take an hour (such as 4.00 p.m.) and find the *row* labelled by that time. Go along that row till you reach the column with your latitude (or a number close to your latitude) at the top. The number that you get to is the angle x that you need to measure to give you the hour line corresponding to that time and latitude. The angle is given in *degrees*. For example, if your latitude is 40° and you want to construct a line for a time of 4.00 p.m., the corresponding value from the tables is

$$x = 48.1^\circ.$$

Flushed with the success of having constructed the base of the sundial you can now finish off the sundial proper. To do this you need to stick a gnomon onto the base. This should be a pointer aligned along the noon line pointing from the centre of the base and it should be at the angle ϕ of your latitude and point due north. To keep the pointer fixed at the right angle it is conventional to use some artistic design such as a reclining Cherub, but this is optional for your own personal sundial. A simpler solution is to cut out a right angle triangle with an angle ϕ at the base. You should then stick this on to the horizontal base (possibly by cutting a slot in the base and sticking it through the slot). To align the sundial, place it on a horizontal table and turn it round until the pointer points to north, just as you did for the Equatorial sundial. The resulting sundial will then look something like Figure 3.11.

3.6 The vertical sundial

The vertical sundial is just like the horizontal sundial except that the dial base is vertical. Typically a vertical sundial is attached to a wall. Many old buildings have vertical sundials, for example Christ's College, Cambridge, and Brasenose College, Oxford, both have fine examples. The design of the sundial depends on which way the

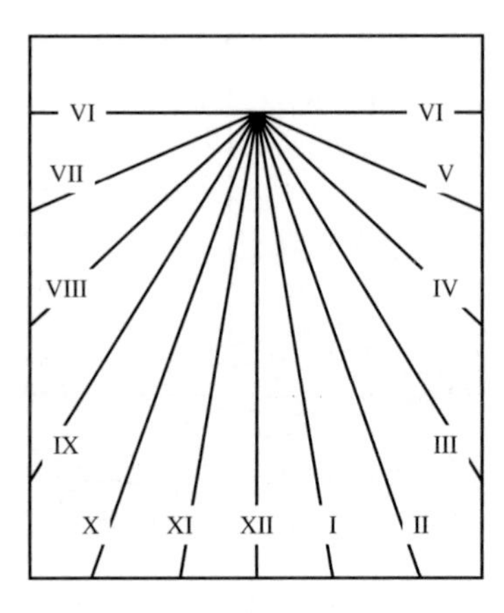

Figure 3.12: The face of a south-facing vertical dial.

wall is facing, but the best sundials are on walls which face due south, firstly because they will get more sunlight that way and will therefore work for longer during the day, and second because they are easier to design. A big difference between a vertical and a horizontal sundial is that the shadow on a vertical sundial goes round *anticlockwise* rather than clockwise. The line on the wall of a vertical sundial corresponding to 12.00 noon is at the *bottom* of the dial and the line for 1.00 p.m. is measured anticlockwise from this.

A typical example of the face of such a vertical sundial is given in Figure 3.12 and a vertical south-facing dial in use can be seen in Figure 3.9.

To construct the dial of a vertical sundial you can again use tables or a formula. To use *tables* you use the same table as for the horizontal sundial. Again you find the row in the tables which corresponds to the time, but this time go along to the column which has your latitude at the *bottom*. For example, if you want the hour angle x for a time 4.00 p.m. at a latitude of 40° then the corresponding angle is 53°. In a vertical sundial this should be measured *anticlockwise* from the noon line.

Instead of using tables you can again use a formula. In this case the formula is

$$x = \tan^{-1}\left[\tan(15T)\cos(\phi)\right].$$

Exercise. *If* $T = 2.30$ *p.m. and you are in New York, use this formula and Table 3.1 to find* x.

In the further problems 2(iv) you are asked to derive this formula for yourself. It is possible to work out similar formulae for walls at any angle, but these are a lot more complicated.

Suppose that you have constructed the base of the vertical dial. The pointer should be attached to the base just as for a horizontal sundial, but in this case it should make an angle of $90^\circ - \phi$ to the vertical. You can use a right angle triangle in the same way as for the horizontal sundial. Having done this you can mount the sundial on a vertical south-facing wall for all to admire.

3.7 The analemmatic sundial

The analemmatic sundial is very different from most people's concept of a sundial. But it has two advantages over a usual sundial, especially when used as part of a school project.

Firstly, it is easy to build a large analemmatic sundial which when set up looks extremely impressive. Secondly an analemmatic sundial can use a human being as the pointer. It is thus *much more fun* than a conventional sundial. The main disadvantage (or perhaps it is really an advantage) of an analemmatic sundial is that it requires rather more mathematics to construct one than the previous sundials, and certainly far more mathematics to understand the principles behind it. We will leave out the principles here but you can find it in the references we give at the end.

The basic design of an analemmatic sundial is illustrated in Figure 3.13. The differences between it and a conventional sundial are obvious. Firstly, the dial is a horizontal *ellipse*. Secondly the pointer is *vertical* and, as remarked earlier, can easily be *you*. Thirdly, and most importantly, the position of the pointer *changes according to the declination of the sun and hence the date on which the sundial is used.* If a human being is the pointer then they must stand on a mark on the ground lablled by the date. These marks are calculated by using the equation for the declination of the sun. A large analemmatic sundial was completed recently at St Mary's school at Calne, Wiltshire, and you can see a sixth form student standing on the dial in Figure 3.14.

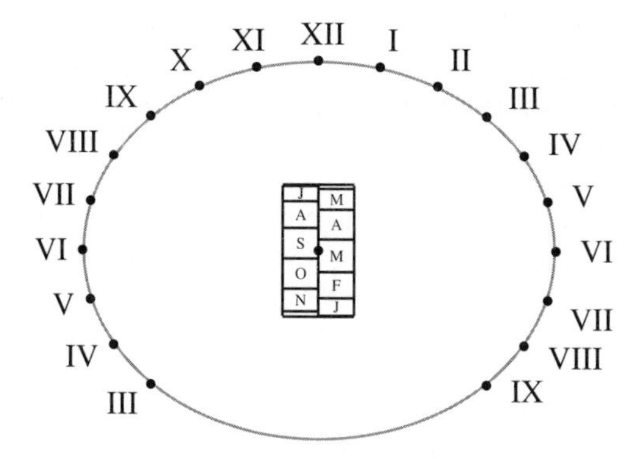

Figure 3.13: An analemmatic sundial.

Figure 3.14: An analemmatic sundial at St Mary's school, Calne, Wiltshire.

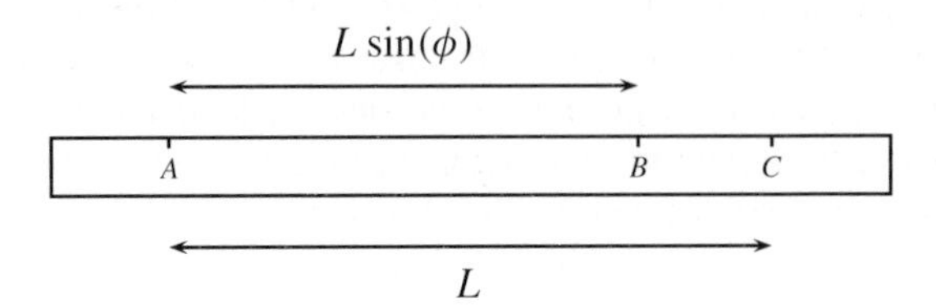

Figure 3.15: A ruler for drawing an ellipse.

To start the construction of an analemmatic sundial you need to draw an *ellipse* which is a curve rather like a squashed circle. It is possible to draw an ellipse using two fixed points and a length of string tied between them. This works well for a larger ellipse but is rather inaccurate for a smaller one. A better way to draw an ellipse is to draw a cross as shown in Figure 3.16. It has two lines at right angles, one running from $S(outh)$ to $N(orth)$ and another from $W(est)$ to $E(ast)$. These cross at the point marked O. To draw an ellipse (or analemmatic sundial) of width $2L$ from W–E, take a ruler and mark two points A and C on it a distance L apart. Now mark a further point B at a distance $L\sin(\phi)$ from A (remembering that ϕ is your latitude). Your ruler should look like Figure 3.15.

Move the ruler around keeping the point B on the line W–E and the point C on the line S–N. Mark where the point A occurs as you move the ruler around. The marks

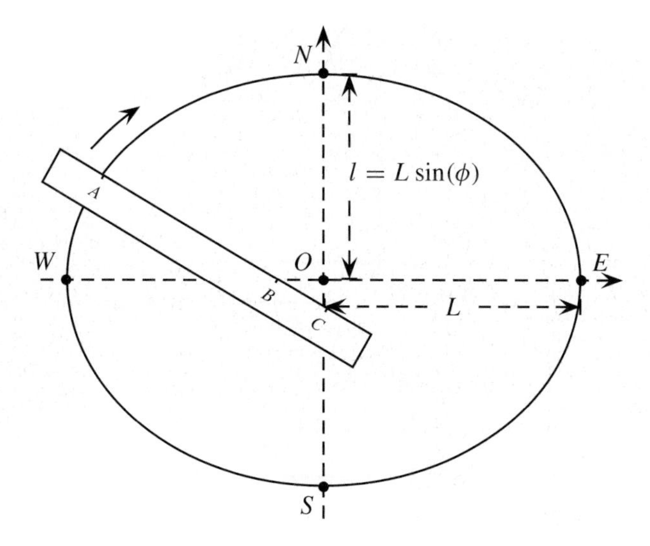

Figure 3.16: Drawing an ellipse.

that you draw then fall on an ellipse. The two lines on the cross that you have drawn are called the *principal axes of the ellipse*. The (shorter) line S–N of length $2L\sin(\phi)$ is called the *minor axis* and the (longer) line W–E of length $2L$ the *major axis*. Set up the analemmatic sundial so that the line S–N points south–north.

You now need to draw a dial. In the earlier sundials this was a series of *lines*. On the analemmatic sundial, the markings are a series of *points* at positions (X, Y) on the ellipse, with each hour corresponding to a point. To locate the point for an hour T after noon multiply T by 15°. Using your calculator calculate X where

$$\boxed{X = L\sin(15T).}$$

For an ellipse with $L = 30\,\text{cm}$ and a time of 3.00 p.m., $T = 45°$ and this gives $X = 30\sin(45°) = 21.2\,\text{cm}$. Now measure a distance X from O to the *right* along the line W–E. Draw a line at right angles to W–E from this point in the direction of N if T is before 6.00 p.m. and S if T is after 6.00 p.m. Where this line crosses the ellipse mark the point for the hour with a conveniently artistic label. In fact the distance Y of this point from the line W–E is given by

$$\boxed{Y = L\sin(\phi)\cos(15T)}$$

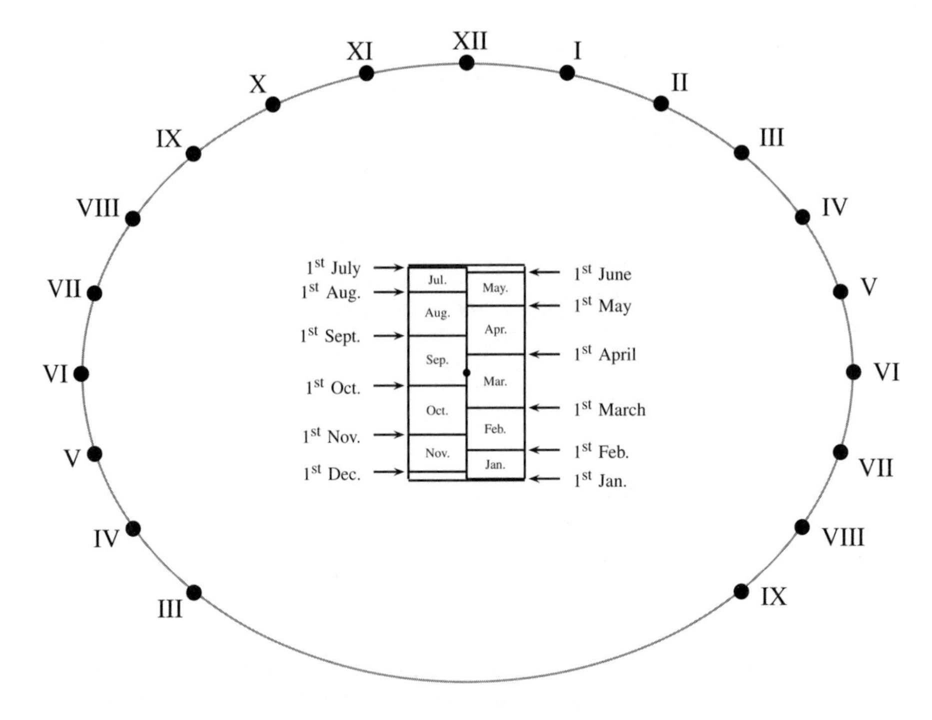

Figure 3.17: The completed dial for Bath, England.

and you can lay out the dial by computing the points X and Y for different times T and then plotting them as you would plot points on a graph. For hours *before* noon, count the number of hours until noon, work out X as before, and this time plot the hour point to the *left* of O. If you do this for all the hours then you will get a complete dial as illustrated in Figure 3.17. Now we come to the location of the pointer which is placed along the line S–N at a position P which depends upon the month. This position is calculated from the declination Z of the sun and is given by the formula

$$P = L \tan(Z) \cos(\phi).$$

To use this formula, take the values of Z from Table 3.4 and then using your calculator work out P. You can then mark a series of points on a scale corresponding to different dates. If this is for a sundial where a person is the pointer, make sure that it is clear where they should stand. The completed dial, with scale, will then look something like Figure 3.17.

Place this scale along the line S–N with the mid-point (which is where the two equinoxes occur) at O, with June close to N and December close to S.

Table 3.4: The declination of the sun.

Date	Z
Jan 1st	−23.1
Feb 1st	−17.3
Mar 1st	−7.8
Apr 1st	4.3
May 1st	14.9
June 1st	22.0
June 21st	23.4
July 1st	23.2
Aug 1st	18.2
Sep 1st	8.5
Oct 1st	−3.0
Nov 1st	−14.2
Dec 1st	−21.7
Dec 21st	−23.4

To use the sundial in, say, August place a *vertical* pointer on the north–south (centre) line in the region marked Aug. at a point which corresponds (approximately) to the date. So for example at the start of August towards the top of this interval. If you are casting the shadow on a big dial in a garden say, you must stand above this point on the centre line. From seeing where the shadow cast by the pointer meets an hour point, we can tell the time.

There is *no* restriction to the size of the analemmatic sundial, however for one that will work well with a human pointer we would recommend that $L = 1$ m. When constructing such a sundial, make sure that it is on level ground away from any trees which might cast unwanted shadows.

A summary of all the measurements for an analemmatic dial is shown in Figure 3.18.

3.8 What time does a sundial show?

If you have ever compared the time told by a sundial in a local garden with the time as measured with a watch you will very likely find that they are not in agreement. There are several possible reasons for this. It could be that the sundial has simply been set up incorrectly, because it is vital for the accuracy of an Equatorial or a horizontal sundial that the shadow-casting object points directly toward Polaris. To do this you have to be careful in pointing the base of the sundial so that it is aligned north–south, and ensuring that the base of a horizontal sundial is indeed horizontal. You must also be careful in drawing the dial as accurately as you can; indeed, many ornamental sundials have dials drawn like clock faces and as we have seen, this leads to an inaccurate reading of the time.

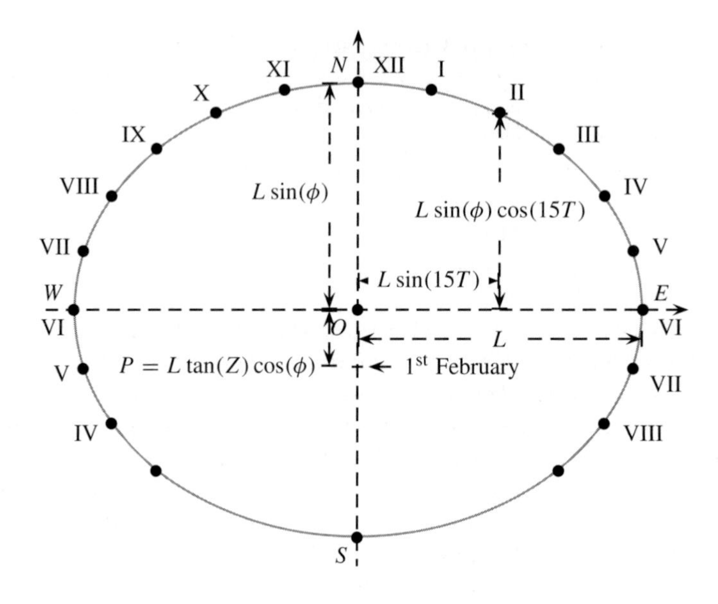

Figure 3.18: A summary of the construction of an analemmatic sundial.

Let's suppose that all errors of construction have been eliminated. The sundial is still unlikely to show the same time as your watch. There are three main sources of this difference. They are:

1. the difference between time in summer and winter;
2. the difference in time around the world;
3. the change in the length of the day during the year.

By using some mathematics we can calculate the value of this difference so that the time, as measured on the sundial, can be converted to the time as measured by a watch. We will now look at each of these differences in turn.

Summer and winter

During the First World War, maximum use had to be made of the available daylight, which is shorter in the winter than in the summer. To achieve this it was decided that the time shown on the clocks in the UK should change at dates close to the spring and autumn equinoxes. In the summer the clocks should go forward one hour so that at noon the clocks should read 1.00 p.m. This new time was called British Summer Time (BST). In the winter the clocks go back one hour to read 12.00 at noon in Greenwich Mean Time (GMT). Many other countries also change the time on the clocks between winter and summer. For example, in North America this is called *Daylight Saving Time*. (An attempt to keep the clocks always at BST was tried in the 1960s but was not a success.) If your sundial is designed to read 12.00 at noon then it will show an extra one hour difference from your watch during the summer months. This change is easy to allow for, and to change to summer time you must add one hour to the time measured on the sundial. Alternatively, a lot of sundials set 1.00 p.m. to be the noon line as they are more often used during the summer when the sun is more likely to be shining.

Time around the world

It takes the sun 24 hours to go round the world. Thus in one hour it travels $360°/24 = 15°$ or it takes four minutes to travel one degree. This means that if you are one degree *west* of Greenwich then noon at your location will occur four minutes after noon occurs in Greenwich. At any particular point we can define *local* time by setting our watches to be at 12.00 when the sun is at its highest point in the sky. Indeed, until the nineteenth century, this was how time was measured, so that different towns had different times. Bristol is about 2.5° west of Greenwich so that time in Bristol was always 10 minutes behind time in London. (Things always do go a bit more slowly in the West Country.) This worked fine until the coming of the railways and the need to keep to timetables. It rapidly became impossible to use local time and instead everyone in the UK fixed their clocks by the time at Greenwich. The bottom line of this argument is that a sundial in Bristol *runs slow* and needs to have a correction of 10 minutes added to it, so that if a sundial reads 3.00 p.m. in Bristol, it is really 3.10 p.m. in Greenwich. In general a correction is needed for every town in the UK. To find this, find out your *longitude* in degrees. The table of latitude and longitude given in Section 3.2 will help you here. This is how you correct your sundial if you live in the UK.

> Add 4 minutes to your sundial for every degree that you are west of Greenwich.
> Subtract 4 minutes for every degree east of Greenwich.

The whole world is divided in to *time zones* such that each location in a time zone has the same time. Thus the whole of the UK is one time zone and most of the rest of Europe is in a different time zone which is one hour ahead of Greenwich mean time. The United States is divided into four time zones called (from west to east) Pacific, Mountain, Central, and Eastern. Each of these zones has a line of longitude called the meridian running through it. These are given as follows

Time zone	Meridian longitude
Pacific	120°W
Mountain	105°W
Central	90°W
Eastern	75°W

Any place on the meridian longitude will measure noon and 12.00 at the same time. Greenwich itself lies on the Prime Meridian which has longitude of 0°. If you live in a time zone, but not on the meridian longitude, then the time measured on your sundial will differ from that of your time zone. For example, if you live in one of the above time zones in the USA, then you must subtract the meridian longitude from your longitude and multiply by 4, to give the number of minutes that you need to add to your sundial for it to tell the time in your time zone. Other values for meridians and time zones can be found in an atlas.

The equation of time

A third, large, and much harder to estimate, difference between time as measured on your watch, and the time shown on your sundial, arises from the way that the Earth goes around the sun. The Earth's orbit around the sun is an ellipse with the sun at its focus. At certain points of the year the Earth is nearer to the sun and moves *faster* than when it is further away, so that the sun appears to move faster across the sky. Furthermore, the axis of the Earth is tilted at 23.5° to the direction of motion around the orbit and this also affects the time it takes for the sun to appear to move around the Earth. The combination of these two factors means that the time it takes for the sun to go from its highest point in one day to the highest point the next is not exactly 24 hours as measured by a watch. For example in December the length of a day is about 24 hours and 30 seconds which is *longer* than 24 hours. In contrast the day in September is *shorter* than 24 hours. As it is very inconvenient to have an hour which changes from day to day, we define 24 hours to the average length of a day taken over the whole of the year. A clock or a watch is designed to measure this hour and this is called *mean time* (hence the name Greenwich Mean Time). The (average) second is now defined more precisely as the time it takes for one Ceasium 133 atom to oscillate 9192 631 770 times.

Now, suppose that we set our sundial and our watch to be both at exactly at 12.00 noon on Christmas Day (December 25th). On December 26th the sundial will register noon (i.e., the shadow will fall on the noon line) at 12.00 noon plus 30 seconds. This may not sound very much, but over a month this difference increases so that by the beginning of February the sundial is nearly 15 minutes slow, registering noon (so the sundial shows 12.00) at about 12.15 p.m. when compared to your watch. This is quite a big difference if you think about it. You would probably miss your bus or train if you were to rely on such a sundial to tell you the time. As the date changes the difference in the time measured between the sundial and the watch changes. On some days there is no difference at all. These days vary slightly on leap years, but on average they are: April 16th, June 14th, September 2nd, and December 25th. The difference between the time measured by a clock and the time measured by a sundial is called the *equation of time* or E(date). It can be calculated exactly but this calculation is rather difficult. If you are interested, a scientific account together with a computer program (in Matlab) to calculate and plot E(date) is given in the chapter on sundials by Oettli and Schilt in the book *Solving Problems in Scientific Computing using Maple and Matlab* by

Table 3.5: Monthly values of the equation of time.

Month	E (minutes)
January	9
February	14
March	9
April	0
May	−3
June	0
July	5
August	5
September	−5
October	−14
November	−15
December	−5

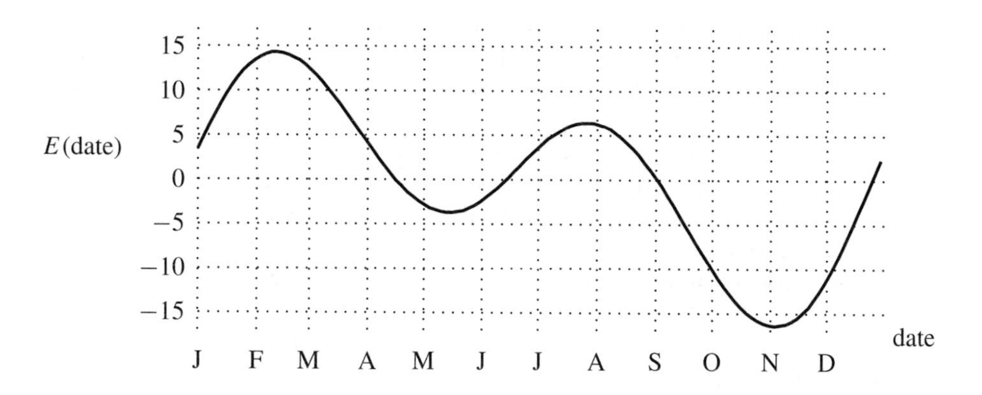

Figure 3.19: The equation of time.

W. Ganger and J. Hrebicek. Table 3.5 shows the average value of E(date) in minutes for the different months of the year. (Although be warned, E(date) can change quite a lot during a month.) In this table a negative value of E(date) means that the sundial is running fast.

A graph of E(date) is shown in Figure 3.19 and is simple to use. Take the reading from the sundial and then *add* the value on the graph for today. Remember, you will also need to correct for your longitude and possibly summer time.

Exercise. *In Bristol in mid-November a sundial, set up to register 12.00 at noon, says it is 3.00 p.m. Show that the time as measured on a watch is 2.55 p.m. When will this sundial show 4.00 p.m. on July 1st?*

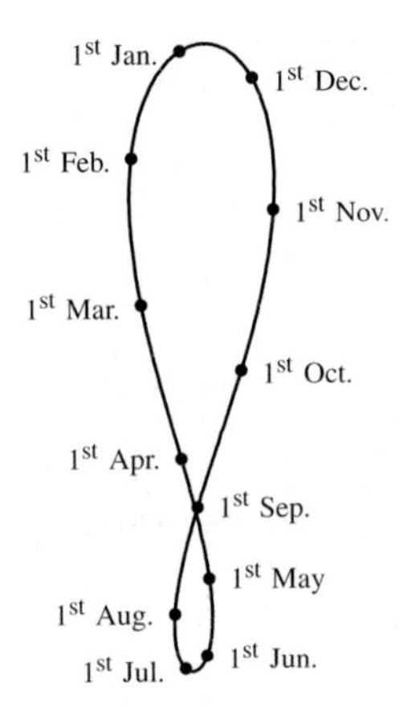

Figure 3.20: An analemma.

A good way to think of the equation of time is as follows. Suppose that you were to stick a vertical post into the ground and make a mark showing the top of the shadow cast by the sun at 12.00 noon, as measured by a clock, each day for a year (making an allowance for the change from winter time to summer time, so that in the UK this mark should always be made at 12.00 GMT). If each day lasted precisely 24 hours then the marks that you make would lie on a straight line going through north–south. Instead they lie on an elegant figure of eight curve called an *analemma* which crosses the north–south line on the dates December 25th, etc., when $E(\text{date})$ is zero. An analemma is illustrated in Figure 3.20. In the exercises we encourage you to record your own analemma over a year of observations.

Adjusting a sundial

Suppose now that you have made a sundial, how can you convert from the time it tells to the time shown on your watch? There are two ways of doing this: either by doing some calculations, or by modifying the design.

Making some calculations

If you have constructed a standard sundial then you need make the three corrections discussed above. To correct for summer time just add an hour. To correct for your longitude add or subtract the appropriate number of minutes as calculated in Section 3.8. Now, to correct for the equation of time, look up your date on the tables or the graph in the last section. If your sundial reads a time T (say 3.00 p.m.) then *add* the values of $E(\text{date})$ to T, so that if the sundial reads 3.00 p.m. in February (on average) the true time (on a clock) is 3.15 p.m. and in May (on average) the true time is 2.55 p.m. You can stick a graph of the equation of time next to your sundial to make this calculation easier.

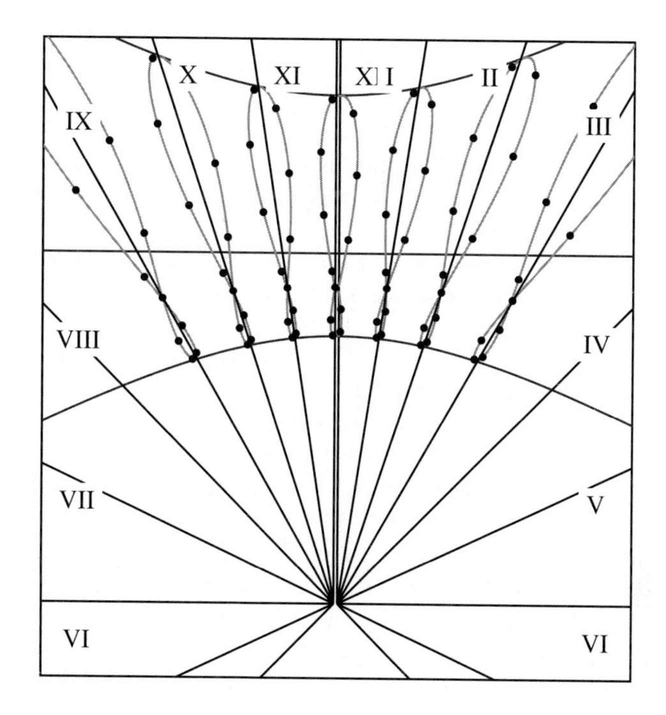

Figure 3.21: The base of a horizontal sundial at 35.6°N (Tokyo).

Changing the design
It is possible to make some design changes to your sundial to do these calculations automatically. To make a sundial which works in the summer add one hour to each of the hour lines. To correct it for your longitude, shift each of the hour lines by the appropriate number of minutes. For example, in Bristol, the sundial runs 10 minutes slow so that if you want to find the hour line for 3.00 p.m. GMT then you should put this at the place where the hour line for 2.50 p.m. would normally lie. More generally, in the formulae for the hour angle (for either the Equatorial, horizontal, vertical, or analemmatic dials) instead of setting T hours, instead set T hours minus 10 minutes (or whatever is the correction for your own location). To correct for the equation of time is harder. A very elegant solution is to replace each hour line by an analemma. An example of such a dial for a horizontal sundial is given in Figure 3.21.

In this design, to find when it is 3.00 p.m. you wait until the shadow cast by the end of the pointer goes through the analemma drawn on the 3.00 p.m. time line at the point corresponding to the day's date. To help you estimate where on each analemma each day falls, the first day of each month has been drawn as a larger dot. Although this is more complicated than using a simple hour line, you will be rewarded by the knowledge that you have found the time as accurately as you possibly can.

3.9 Exercises

First session

1. Think of as many ways to tell the time as you can.
2. Suppose that you are on a ship off the coast of Canada on a clear night. How could you find out your latitude?
3. Suppose that you live at the North Pole (chilly!). What is your latitude? Work out the angle of the sun at noon during mid-winter. Your answer will be surprising. What do you think it means?
4. The Tropic of Cancer has latitude 23.5°. Show that at noon during mid-summer the sun will be directly overhead. How long is the shadow cast by the sun at this time and place?
5. (a) Can you think of a shape which casts a triangular shadow in one direction and a square shadow in another?
   (b) For a harder problem, find a collection of shapes which casts the shadow MORNING in the morning and AFTERNOON in the afternoon? Hint: Think of a Venetian blind. Can you extend this idea?
6. If you are feeling creative then experiment with shadows by making some animal shadows with your hands.

Second session

Some of these problems are constructional. If it is a sunny day then you should test your sundials outside, lining them up as appropriate. (It helps to have a compass to align them with north.)

1. Photocopy the sundial given in Figure 3.22 onto a paper sheet. Cut it out and fold it as indicated. Then stick it onto a base to make a paper sundial that will work at a latitude of 52°.
2. Draw your own horizontal sundial for a chosen latitude using either

   (a) the formula or
   (b) the given tables.

   Paste this onto card. Cut out a pointer by drawing a right angled triangle with the angle of your latitude and put some tabs onto the bottom. Now glue it onto the base to make a horizontal sundial.
3. Following a similar method make a vertical sundial for a south-facing wall (assuming that one is available).
4. Find the angle of the 6.00 p.m. mark for different latitudes. You should find that it is always at 90° to the noon mark irrespective of your latitude. Can you work out why?
5. Draw a disk for an Equatorial sundial onto card, with a reverse disk on the bottom. Stick a wooden skewer through the centre and adjust the length as described in the text so that the disk and needle make an angle of your latitude with the table.

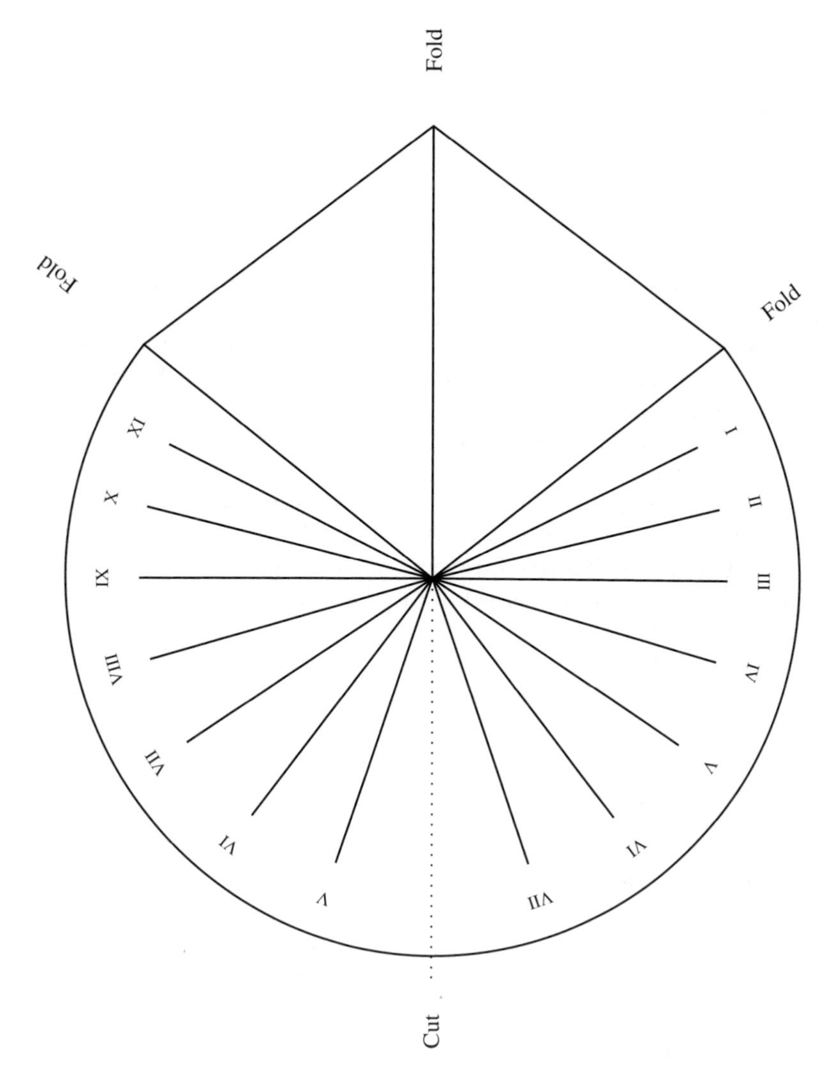

Figure 3.22: A cut out and keep sundial for 52°N.

Add a bit of Blu-tack and you have an Equatorial sundial which uses the top of the disk in summer and the bottom in winter.

6. Imagine you are in Cardiff and a sundial shows 11.25 a.m. on May 1st. What time does your watch show?
7. Find your longitude and work out the correction that you need to make so that your sundial will tell the correct time in your time zone.
8. Would a horizontal sundial work at the equator?
9. What would a sundial look like in Australia?

For a longer-term project, construct an analemmatic sundial for your latitude. Make this large enough so that a person can be the pointer. With imagination you can make a very nice sundial with suitably chosen ways of making the numbers look impressive. Maybe you could incorporate the equation of time into the design.

Field trips and projects

There are many interesting sundial projects and here we suggest a few.

(i) Find a fixed object in a clear open space. A vertical stick, 1m or so is ideal. Plot the points made by the end of the shadow during the day. These points should lie on a curve called a hyperbola. The hyperbola will depend on the declination of the sun and so will vary, and at the two equinoxes will lie on a *straight line*.

(ii) Choose one of the sundials we describe and make it. This could be a joint project with art.

(iii) Most computers contain a program called a *spreadsheet*. The standard spreadsheet provided by Microsoft is called `Excel`. Spreadsheets are ideal for performing simple but repetitive calculations. A good, but quite challenging project is to write a spreadsheet to calculate the hour angles on your sundial using the various formulae given here.

(iv) This project takes a whole year! As often as possible, every day if you can, record the position of the shadow cast by the sun of some fixed object at 12.00 at each day in the year. If you do this then you should find that you get an analemma, such as that shown in Figure 3.20. In future years you can use this analemma to give an accurate time for 12.00 noon.

For an even longer-term project, record how the analemma changes from one year to the next. Here you will see the influence of leap years.

(v) Good sundials of many different designs can be found all around the world. Many examples are given in the books listed in the references. It is well worth trying to find some in your area and paying them a visit.

3.10 Further problems

1. Finding the declination of the sun

In this first problem we will see how to derive the formulae that give you the angle of the sun at mid-summer and at mid-winter in the northern hemisphere.

To a good approximation the world is a sphere. This sphere has an axis around which it spins which is at 90° to the plane of the Equator. This plane is itself at an angle of 23.5° to the plane in which the Earth orbits the sun. At midday on mid-summer the axis (at the North Pole) points *towards* the sun. At mid-winter it points *away* from the sun. The sun itself can be considered to be so large and so far away from the Earth that its rays all strike the whole of the Earth as parallel lines in the same direction as the orbital plane.

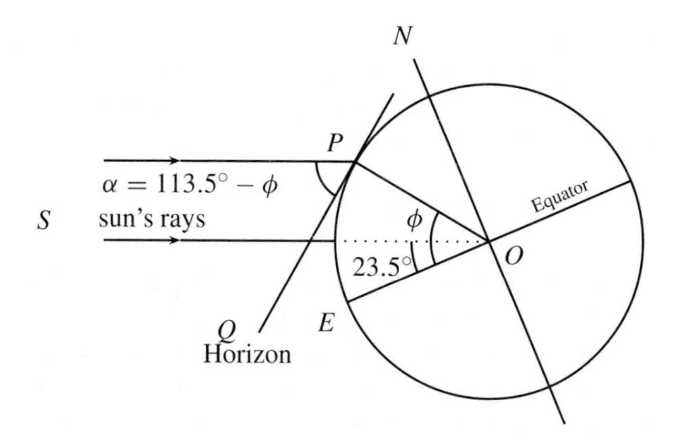

Figure 3.23: Mid-summer.

If you live at a latitude of ϕ then you are at a point which is at an angle ϕ measured with respect to the Earth's Equator, and at midday this point is directly in the path of the sun's rays. The points which make up your horizon (that is a line which corresponds to what you think of as horizontal) lie on a line which is a tangent to the sphere at the point where you are. This line is at 90° to the radius joining you to the Earth's centre. What we are interested in finding out is the angle α between the sun's rays and this line marking the horizontal.

(i) Figure 3.23 summarizes the above situation. In this picture the centre of the Earth is at O and you are situated at the point P at noon on the northern hemisphere during mid-summer. This point has a latitude of ϕ. At this time of year the Earth's axis points *towards* the sun. The sun's rays are all parallel to the line OS which is at an angle of 23.5° to the Equator OE. Check that you understand how this diagram relates to the above discussion.

(ii) The angle α can be worked out using some simple properties of triangles and parallel lines.

(a) Show that the line OP makes an angle of $\phi - 23.5^\circ$ with the line OS.

(b) Draw a line PQ which corresponds to a horizontal line at P. This line is at 90° to the line OP. Using the fact that the angles in a triangle add up to 180°, show that the horizontal line PQ makes an angle of $90^\circ + 23.5^\circ - \phi = 113.5^\circ - \phi$ with the line OS.

(c) From properties of parallel lines, it follows that if we draw a line at P which is parallel to OS and which is the line that a ray from the sun would make at P then this line also makes an angle of $113.5^\circ - \phi$ with the line PQ. But this is the angle α. We deduce that

$$\alpha = 113.5^\circ - \phi.$$

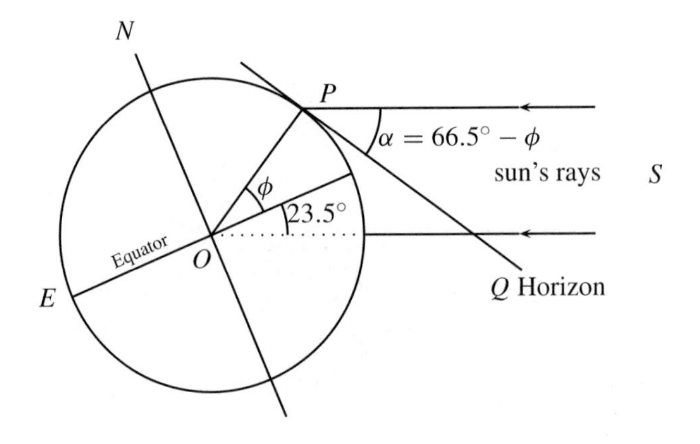

Figure 3.24: Mid-winter.

(iii) During mid-winter the Earth's axis points away from the sun at midday. Check that you can see how this corresponds to the situation shown in Figure 3.24.

(iv) By drawing a similar triangle to the last one, deduce that

$$\alpha = 90^\circ - 23.5^\circ - \phi = 66.5^\circ - \phi.$$

(v) You can in principle repeat this calculation to work out the angle of the sun at noon for any date during the year to work out the declination Z. However this calculation uses much more complicated geometry than we have discussed here.

(vi) Using these diagrams show that the angle Polaris makes with the horizon is ϕ.

2. Plotting the horizontal dial

In this problem we will derive the formula which gives the base of the horizontal sundial. This calculation involves some trigonometry, and is rather harder than that in problem 1.

We can think of the dial of a horizontal sundial as lying underneath the dial for an Equatorial sundial. The Equatorial dial is at an angle of $90^\circ - \phi$ relative to the horizontal dial with the pointer passing through both dials. On the Equatorial dial the lines marking the hours are at constant angles of $15T$ where T is the desired time in hours. What we need to do is to project these lines down onto the horizontal dial.

(i) Suppose that the pointer passes through the horizontal dial at the point O and then through the equatorial dial at the point P. The equatorial dial is perpendicular to the pointer. The point where the noon line of the equatorial dial meets the noon line of the horizontal dial is N. Check that you understand the diagram in Figure 3.25.

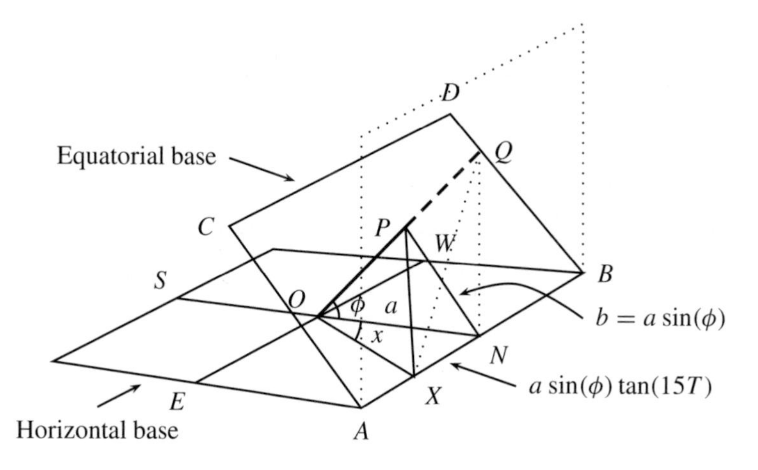

Figure 3.25: Calculating the angles for a horizontal dial.

(a) Show that OPN is a right angle triangle with angle ϕ at O and angle $90° - \phi$ at N.

(b) Hence show that if the line ON has length a and PN has length b then

$$b = a\sin(\phi).$$

(ii) Now draw a line on both the horizontal and equatorial dials which is at right angles to the noon line ON. If X is a point on this line and if a shadow on the Equatorial dial is given by the line PX then the resulting shadow on the horizontal dial is the line OX. If T is the time then the angle XPN is $15T$.

(a) Show that the length c of the line NX is given by

$$c = b\tan(15T) = a\tan(15T)\sin(\phi)$$

(b) Hence show that if x is the angle XON then

$$\tan(x) = c/a = \tan(15T)\sin(\phi).$$

(c) We deduce that

$$x = \tan^{-1}(\tan(15T)\sin(\phi)).$$

(iii) We can summarize the above argument with Figure 3.26 which also allows you to calculate the angle XON graphically (without having to use a calculator). To understand this figure you might like to think of Figure 3.25 being hinged along the line AB. The Equatorial and horizontal dials are unfolded to lie in the same plane. Next the pointer is hinged along ON and also laid flat.

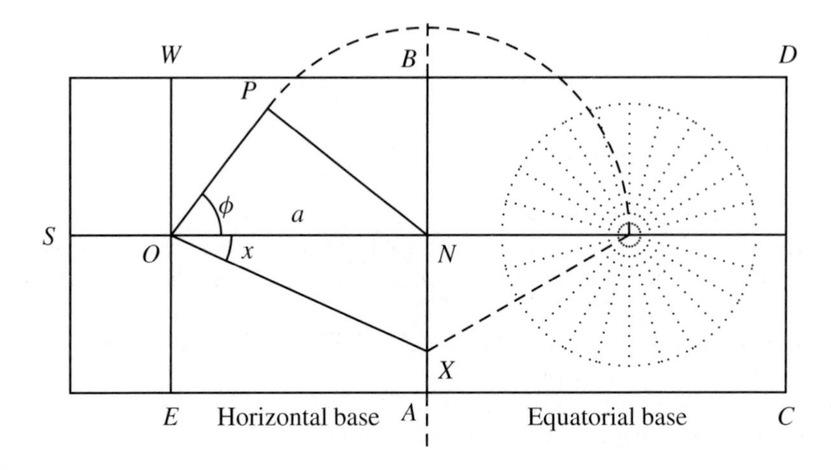

Figure 3.26: A graphic method for drawing a horizontal sundial.

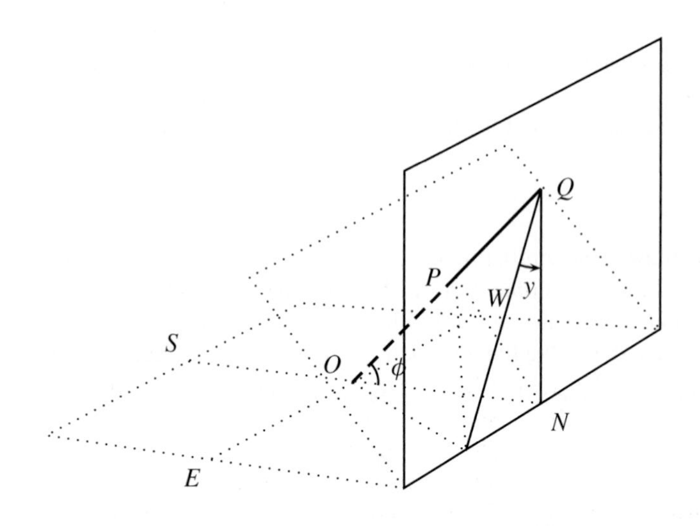

Figure 3.27: Calculating the angles for a vertical dial.

(iv) See if you can copy this argument to obtain the formula

$$y = \tan^{-1}(\tan(15T)\cos(\phi))$$

for the vertical sundial. It might help to compare Figure 3.25 for the horizontal sundial with Figure 3.27 for a vertical one.

3.11 Answers

First session

2. In the northern hemisphere, you find your latitude by measuring the angle between Polaris and the horizon.
3. At the North pole your latitude is 90°. During mid-winter the angle of the sun at noon is $66.5^\circ - 90^\circ = -23.5^\circ$. This answer is negative. During mid-winter the sun is below the horizon and it is dark all day. In contrast, during mid-summer it never gets dark.
4. In mid-summer the angle of the sun is $113.5^\circ - 23.5^\circ = 90^\circ$. Thus the sun is directly overhead and casts no shadow.
5. Two possible answers to (a) are a square based pyramid and a triangular prism (a Toblerone packet). To answer (b) think of a Venetian blind with the blinds angled in a south-easterly direction. In the morning this allows the sun's light to pass through without obstruction. Put a collection of shapes (wooden letters will do) to cast a shadow of MORNING in the path of the rays. In the afternoon these shapes will receive no light from the sun. Next to this blind, construct another blind angled in a south-westerly direction and use this to help cast the shadow of AFTERNOON. With long enough blinds it is in principle possible to produce a sundial which will cast digital shadows for each hour of the day.

Second session

4. If $T = 6$ then $15T = 90^\circ$. Now, $\tan(90) = \infty$ so that $\tan(90)\sin(\phi) = \infty$ for all angles ϕ between 0 and 90°. Thus $\tan^{-1}(\tan(90)\sin(\phi)) = 90^\circ$ regardless of the value of ϕ.
6. Cardiff is 3.2° west so we must add $3.2 \times 4 \approx 13$ mins. to correct for longitude. The equation of time from Table 3.5 shows -3 mins. We add one hour for summer time so that a watch will show 12.35 p.m.
8. At the Equator the pointer of a horizontal sundial is horizontal and lies on the base. It can thus cast no shadow and the sundial will not work.
9. In Australia, the sundial is marked out in the same way but with the big difference that the dial of an Equatorial or a horizontal sundial should be marked out anticlockwise.

3.12 Mathematical notes

A question of degree

In this chapter we have adopted the convention that angles should be expressed in degrees and decimal fractions of degrees. This is consistent with the way that angles are expressed by most calculators. We have avoided radians because most of the young people attending masterclasses are not familiar with them and protractors do not use them. (Radians are usually the convention for angles on a computer, so that any computations on a computer will have to be converted from radians to degrees.) Another notation which we have not used is to express angles in terms of degrees, minutes, and seconds. In this notation one minute is 1/60th of a degree and one second is 1/60th of a minute. For example, an angle of 23.5° would be given as $23^\circ 30'$. In older books on sundials, angles are always expressed in this manner and most of the tables giving the declination of the sun and the angles of a horizontal dial use this convention. It is also widely used in the expression of latitude and longitude. This should be borne in mind if you are interested in following up any of this material by going through the books listed in the references. We believe that it is rather inconvenient to use the degrees, minutes and seconds notation for calculation, and that given that most of the calculations in this chapter will be performed by using a calculator, it is easiest and simplest to stick to decimal fractions of degrees.

The equation of time

The formulae for the equation of time are more complex than those given in the rest of this chapter, but can still be derived by using simple trigonometry. They would make a useful advanced project. Details of their derivation are given in W. Ganger and J. Hrebicek, *Solving Problems in Scientific Computing using Maple and Matlab*. As an illustration of the interrelation between the equation of time and the orbit of the Earth, it is interesting to mark the precise location of the sun on one day over a period of several years. Choose a day when the difference between a solar day and 24 hours is greatest. Christmas Day is a good example. If you do this then you will see a slow drift from one year to the next which is almost exactly corrected every fourth year when we make the correction of the leap year.

3.13 References

There are many good books about sundials, some of which go quite deeply into the mathematics, and others have a lot of information about how to construct a sundial. Here is a small selection, but you should be able to find many others.

- Waugh, A. E. (1973) *Sundials, their Theory and Construction.* Dover.
  We would seriously recommend this book to anyone with an interest in sundials as it gives very detailed instructions on their history and how to make them. Note, however, that it was written before the invention of calculators.
- Jenkins G., and Bear, M. (1989) *Sundials and Timedials.* Tarquin.
  This book is a comprehensive set of cut-out and fold designs for sundials of many different sorts. It also has a nice booklet explaining the theory behind sundials.
- Ganger, W. and Hrebicek, J. (1991). *Solving Problems in Scientific Computing using Maple and Matlab.* Springer.

This book has a chapter on sundials and some excellent computer code to produce the base for a sundial and an analemma automatically as well as calculating the equation of time.

- For more information about why the analemmatic dial works see our online article at `http://plus.maths.org.uk/issue11/`
- Sobel, D. (1996). *Longitude.* Fourth Estate.
  This book is not about sundials. However it gives a brilliant account of the history of finding longitude and the central role that accurately finding time played in this.
- Landes, D. *Revolution in time, clocks and the making of the modern world.* Viking.
  This book takes you on from sundials to show how clocks were invented and the effect that this invention has had on us all.

If you are really interested in learning more about sundials then you can join the British Sundials Society: `http://www.sundials.co.uk/`

4 Magical mathematics

4.1 Introduction

Most of us, if asked, would say that we like to watch magic and magic shows. If you were to think why you like magic you might come up with the following answers:

- magic shows are full of surprises
- magic shows are dramatic
- magic shows are mysterious
- magic shows make you disbelieve the evidence of your own eyes
- magic shows leave you wondering – how does the magician do that?

Another feature of magic that adds to its mystery is that it is really hard to do well, and to put on a good magic show takes a lot of time and practice. There is also no absolute guarantee that all will go well on the night.

Now let's compare mathematics with magic. If we were to ask you why (and let's hope this is true) you like mathematics (at least we assume that you like mathematics enough to be reading this book) then your answers might include the following:

- mathematics is a logical subject
- mathematics has a right answer
- in mathematics you know where you are – there are no hidden surprises
- mathematical truths last for ever
- mathematics has many useful applications.

None of the above is necessarily untrue, however it does lead to a problem in the way that mathematics is perceived. To many people's minds the fact that mathematics is logical seems to imply that it is neither mysterious, nor can it have any surprises. All of which appears to be the opposite of the reasons that we enjoy magic shows. Fortunately, however,

nothing could be further from the truth.

Being logical does not in any way prevent mathematics being both mysterious and full of surprises. In fact as we shall show in this workshop, mathematics lies behind some really good magic tricks. By learning them you will have the opportunity to amaze and impress your friends with your astonishing conjuring skills.

4.2 Magical mathematics

We all think that we know what mathematics is, but if you were asked to define mathematics, then you might find this rather difficult to do. Here is our attempt at a definition.

Start first with the building blocks of mathematics which are abstract quantities such as numbers, shapes, sets, functions, etc. Mathematics is all about finding patterns and links between these. It is finding these patterns, which are called *theorems,* and then *proving that they are true* which is at the heart and soul of mathematics.

The patterns that we find in mathematics are often very remarkable and unexpected and have far-reaching implications.

To see what sort of remarkable patterns we can find and how the above 'definition' of mathematics works in practice we will now give some examples.

Example 1. Triangles are idealized mathematical shapes which have been known about for thousands of years. An important example of a triangle is the right angled triangle which arises naturally when you want to build a rectangular house. Such a triangle has sides of length a, b, and c where we will assume that c is the longest. At first sight there is no relationship or pattern between these three values, indeed, why should there be one?

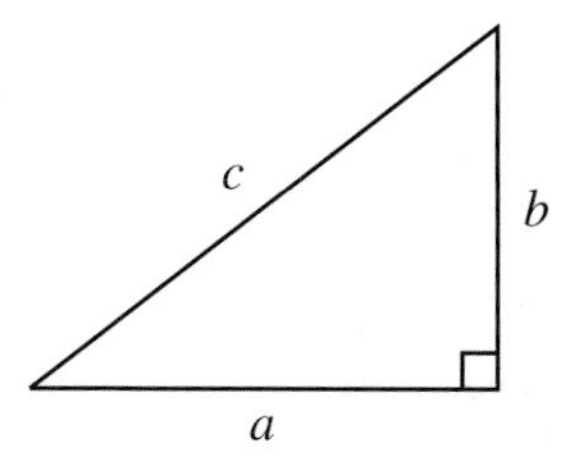

It was the genius of the ancients that found the pattern

$$a^2 + b^2 = c^2$$

which you will recognize as being Pythagoras' theorem. Although this theorem is a direct consequence of a logical argument, it does not stop being a beautiful and surprising result. The remarkable thing about it is, of course, that it applies to *any* right-angled triangle. Herein lies its power and the fact that it has extraordinarily many applications.

Example 2. Another mathematical shape which has been known since antiquity is the circle. Associated with the circle is the number $\pi = 3.141\,592\,653\,589\,793\,2387\ldots$ which is the ratio of the circumference of the (indeed any) circle to its diameter (or alternatively the ratio of the area of the circle to the square of its radius). Now, the number π is naturally associated with the abstract shapes that you find in geometry, and its value was originally calculated by ancient mathematicians such as Archimedes by estimating the area of a circle. However, what is really remarkable is that there are theorems connecting π to many of the other quantities that you find in mathematics. These theorems are beautiful, unexpected, and in every sense magical. Here is one of our favourites which is a particular case of an infinite series called Gregory's formula, discovered in 1671. This links π to the odd numbers and is simply this

$$\pi = 4\left(1 - \frac{1}{3} + \frac{1}{5} - \frac{1}{7} + \frac{1}{9} - \cdots\right).$$

We hope that you can see the pattern here. To calculate π we need to take an *infinite* number of terms in the series. However by taking a finite number of terms in this series you can estimate π to any required level of accuracy. For example the first term gives the estimate

$$\pi \approx 4,$$

the first two terms give the estimate

$$\pi \approx 4\left(1 - \frac{1}{3}\right) = 2.6666\ldots,$$

and the first three terms the estimate

$$\pi \approx 4\left(1 - \frac{1}{3} + \frac{1}{5}\right) = 3.4666\ldots.$$

None of these estimates are particularly good, but they are obviously getting closer to the true value of π, and you can do much better by taking more terms.

Exercise. *Try looking at what happens if you take a large number of terms in the above infinite sum. You can try this out either on a calculator or on a computer. Show by performing some calculations that your answers get closer to π as the number of terms that you take increases.*

There are many other theorems which make remarkable and mysterious links between the number π and other abstract mathematical quantities. Here is one linking π with the square numbers, which was discovered by Euler in 1736:

$$\pi^2 = 6\left(1 + \frac{1}{2^2} + \frac{1}{3^2} + \frac{1}{4^2} + \cdots\right).$$

Exercise. *By taking several terms check that you can approximate π by using this series.*

And finally, the best theorem of all, which was also discovered by Euler and links all the main numbers used in mathematics. These are 1, π, e which is the basis of natural logarithms and which we will meet in Chapter 8, and the *imaginary* number, i which is defined to be the square root of -1, and satisfies $i^2 = -1$. Here is the marvellous link between these four numbers:

$$\boxed{e^{i\pi} + 1 = 0}$$

This result really is truly astonishing, as it takes four quite different ideas in mathematics and links them all together. More than just about any other theorem this result demonstrates that mathematics is truly a unified and surprising, magical, and wonderful subject. As well as being very beautiful, this theorem is also very useful, and plays a central role in helping us to understand the way things change periodically in time. For example, the electricity supply industry, which uses alternating current to supply electricity, uses this theorem and its consequences, every time it designs and operates a power station.

4.3 Mathematical magic

We will now create a few surprises of our own. Theorems as described in the last section hold a great deal of wonder, beauty, and mystery for the mathematician, but, unlike magic shows, are not readily appreciated by the general public. However, because a mathematical theorem deals with abstract quantities, and expresses an often surprising link between them, it is actually quite easy to turn a theorem into a magic trick. What you do is relate the abstract quantities to something that you could use in a magic show, such as a clock or a pack of cards. The theorems then become surprising links between known objects that anyone can appreciate and admire. We will illustrate this by looking at four magical tricks that are all based on mathematics, in particular, subtraction, division, the pigeon-hole principle, and some elementary number theory. As well as emphasizing interesting mathematical properties these tricks all work (they must do!) and are all *great fun to perform.* We suggest that you try them all out and aim to add lots of flourishes to aid the dramatic nature of the performance.

Trick one: Tapping the clock

What is the trick?

To perform this trick you will need a *large* clock face showing the usual numbers 1 to 12. (Do not use a digital clock!) Here is an example of such a clock face

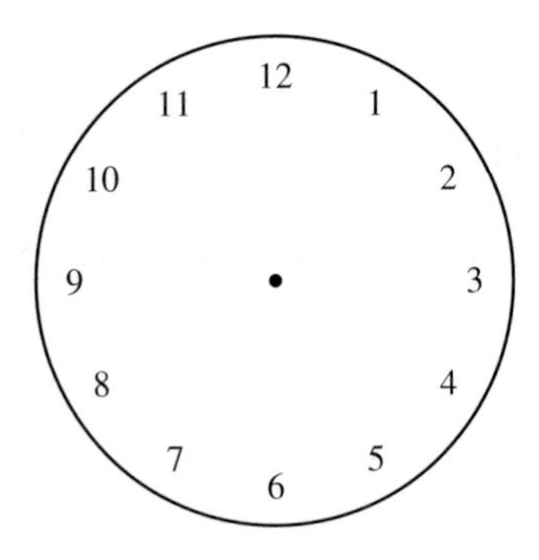

A good way to do this trick is to put a clock face (such as that above) onto an overhead transparency and to use this for the performance. You call for a volunteer from the audience and ask them to choose one of the numbers on the clock face. Ask them to tell someone close to them what that number is (but not to tell you of course). You now start tapping numbers on the clock face apparently at random. Tell your volunteer to start counting (in their head) when you start tapping. They should start the count with their chosen number (say this is the number 8) and to carry on counting every time you tap *until they reach the number 20*. When they reach 20 they should shout loudly to you to STOP. At this point (mysteriously) you will be tapping exactly their chosen number! You will, at this point, be greeted by tremendous applause! If the applause is good enough, then do this trick a few times more.

How does this trick work?

- You do the first eight taps on the clock face *completely at random*. Do this with a flourish. The point of doing this is simply *to confuse your audience*.
- Now make the ninth tap on the number 12 (on the top of the clock face).
- Make all later taps working your way round anticlockwise (so that you tap 12, 11, 10, 9, 8, etc.). With any luck, the audience will be so confused from your first eight random taps, that they won't notice the regularity of the way that you are tapping the clock face in this part of the trick.
- When they shout STOP you will be tapping the right number (as if by magic).

Why does this trick work?

We will now show, mathematically, why this trick works. In fact it is based on the simple mathematical identity, that for any numbers x and y

$$y - (y - x) = x.$$

In fact almost the same identity will be used in some of the magic tricks in the exercises and also later on when we do some “mind reading".

Exercise. *Verify this identity.*

This is an identity relating abstract quantities. To turn it into a magic trick we must relate this result to the numbers on the clock face. Let’s suppose that x is the mystery number chosen by your volunteer. They start counting at this number with every tap that you make until they reach 20. The first question that we ask is *how many taps do you make?* For example if their number is eight then they will count 8, 9, 10, 11, 12, 13, 14, 15, 16, 17, 18, 19, 20. In this case you will have made 13 taps. In general you will make

$21 - x$ taps from start to finish.

Exercise. *Verify this result.*

Now, the first eight of these taps are random taps and play no real role in the trick, other than to deceive your audience. If we subtract these taps from the total, then after the random taps we will have $21 - x - 8 = 13 - x$ taps remaining. We will call the remaining number of taps y so that $y = 13 - x$.

The first of these remaining taps will be on the number 12, then next on the number 11, the third on the number 10, etc. Suppose that starting on the number 12 we make y taps anticlockwise. Where will we stop? If $y = 1$ we will stop at 12, if $y = 2$ we will stop at 11, etc. It is easy to see the general result: in general we stop at the number given by

$$13 - y.$$

But we have already worked out that $y = 13 - x$. Thus the number we stop at is given by

$$13 - (13 - x) = x$$

where we have used the mathematical result we have given above. What have we learnt? Well the mathematics tells us that we stop at x, but of course x is the number chosen by the member of the audience. So we have stopped at exactly the chosen number and the trick has worked.

In the exercises there are several problems based on the ideas in this trick – they are all based on subtraction. Try these out and then see if you can work out some more tricks of your own.

Trick two: Finding the card

This is a very famous trick and in it you will mysteriously find a chosen card in a pack containing 21 cards. It also demonstrates a mathematical result called the *contraction mapping principle* and is based on division by 3.

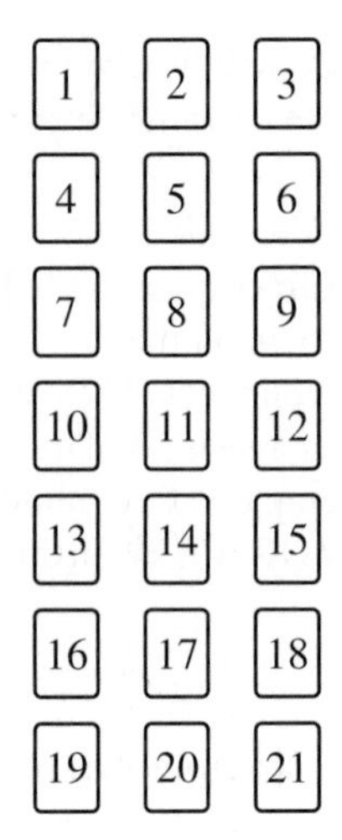

Figure 4.1: Laying out the cards for trick two.

What is the trick?
You count out 21 cards into a pile and give them to a volunteer in the audience. You ask this person to choose a card and to place it anywhere in the pile and to then give the pile back to you. You then deal out the cards, face up, into three columns in the order shown in Figure 4.1.

You ask your volunteer to tell you which of the three columns their card is in (without of course telling you which card it is). You collect up the cards in this column and then place it *between* the other two columns.

You now repeat the process of dealing out the cards in the above order, asking which column the mystery card is in and placing it in the middle. You then perform the dealing process once again.

At this point you are able to tell your volunteer exactly what their card is.

How does this card trick work?
Provided that you have done the trick exactly as described above, the mystery card is the *middle* card in the pack (or card number 11). You can find this card by counting out the pack till you get to the 11th card. This is rather dull and rather gives away the trick. A good method of 'finding' the card is to deal them all out, putting them in a random order on a table, but noting which card is the 11th. When all of the cards have been dealt out point to the mystery card.

Why does this trick work?
As mentioned above, this trick works because of properties of the operation of division by 3. To understand why it works we will say that the *position* of the mystery card in the pile just after it has been placed there by the volunteer is x. For example we could take $x = 6$. Now you deal the cards out into the three columns shown above.

If you look at the card number 6 you will see that it is the second card down from the top of the column it is in. Each of the cards in the original column will end up in one of the three columns after dealing them out and will be y cards down from the top of its column. We have already shown that if $x = 6$ then $y = 2$. Here is the general relationship between x and y:

x	y
1	1
2	1
3	1
4	2
5	2
6	2
7	3
8	3
9	3
10	4
11	4
12	4
13	5
14	5
15	5
16	6
17	6
18	6
19	7
20	7
21	7

If you look at this table you will see that for each value of y there are three corresponding values of x. This is simply because you have dealt the cards into three columns. If you look at the values $x = 3, 6, 9, 12, 15, 18, 21$ you will see that the corresponding values of y are $y = 1, 2, 3, 4, 5, 6, 7$. So it seems that in this case $y = x/3$. Clearly this does not always work as if $x = 1$ then $x/3 = 1/3$ which does not equal the value of $y = 1$. The relationship between x and y can instead be expressed by the following formula

$$y = \lceil x/3 \rceil.$$

Here the symbol $\lceil z \rceil$ means 'take the nearest whole number greater than or equal to z'. So that $\lceil 2 \rceil = 2$ and $\lceil 2.1 \rceil = 3$.

Exercise. *Calculate* $\lceil 5.5 \rceil$. *Is* $\lceil (2.7 + 2.3) \rceil = \lceil 2.7 \rceil + \lceil 2.3 \rceil$?

Now suppose that your volunteer chooses the column which their card x is in and you collect up the cards and place this pile in the middle of the other two piles. What position does x end up in? We can calculate this quite easily from what we have already found out. The card x is y cards from the top of the chosen column. You then put a pile with seven cards on top of this. Thus the new position of the card x is $y + 7$. We can think of this in the following way. The process of dealing out the cards, choosing a column, and then reassembling the three columns performs an operation on the position x of the mystery card to move it to a new position. This new position is given by the function

$$x \to 7 + \lceil x/3 \rceil.$$

By doing this trick we simply apply this operation three times to x. Here is what happens when you do this:

x	First deal	Second deal	Third deal
1	8	10	11
2	8	10	11
3	8	10	11
4	9	10	11
5	9	10	11
6	9	10	11
7	10	11	11
8	10	11	11
9	10	11	11
10	11	11	11
11	11	11	11
12	11	11	11
13	12	11	11
14	12	11	11
15	12	11	11
16	13	12	11
17	13	12	11
18	13	12	11
19	14	12	11
20	14	12	11
21	14	12	11

If you look at this you can see clearly that after three deals all of the cards end up in position 11 in the pack. This is why the trick works. In fact the difference between the position of the card and 11 decreases by a factor of 3 every time that you do the deal. (This is called a contraction principle.) The number 11 is called a *fixed point* of the map of dealing, choosing a column, and putting it into the middle of the other two piles. Once a card has got to the 11th position it stays there for ever more.

Having seen how and why this trick works, it is easy to think about how it might be generalized. For example what happens if you take a number $N = p \times q$ cards and deal them into p piles of q cards each? For the trick to work you will need both p and q to be odd and you always put the chosen column into the middle of the other columns.

Exercise. *Why must p and q both be odd?*

Exercise. *If you have a pack of 49 cards which you deal into seven columns of seven cards, how many times do you need to deal out the cards and then put the chosen column into the middle until the chosen card is at the 25th position in the pack? Try this out for yourselves and you will find that the answer is rather surprising!*

Trick three: The cards that know how to spell

This is a great trick which has endless variations which you can make up for yourselves. It uses a mathematical result called the *pigeon-hole principle*.

What is the trick?

You hold up a pack of 13 cards and tell your audience that these cards are super-intelligent as they know how to spell their own names. In the pile there are the cards Ace, Two, Three, Four, Five, Six, Seven, Eight, Nine, Ten, Jack, Queen, and King. Starting from the top of the pack (which has the cards face down) you take off the first three cards one by one placing each card at the *bottom* of the pack. As you count out each of the three cards you count out the letters A, C, and E. These of course spell ACE. You turn the next card on the top of the pack over and it is the Ace. Place the Ace on the table in front of you. Now count off another three cards, putting them on the bottom of the pile and saying the letters T, W, and O. The next card that you turn over is the Two. Place the Two in front of you next to the Ace. Next count out five cards, placing them on the bottom counting out the letters T,H,R,E, and E. The next card that you turn over and place on the table is the Three. You carry on doing this, spelling out the names of each card and then turing it over. Eventually you will get to the last card in the pack, which is, of course, the King.

How does this trick work?

Before you start the trick, put the 13 cards into the following order

$$3\ 8\ 7\ A\ Q\ 6\ 4\ 2\ J\ K\ 10\ 9\ 5$$

The trick will then work itself!

Why does this trick work?
Imagine that we have 13 boxes in order as illustrated below.

These boxes represent the position of each of our cards in the pile. We want the first card that we turn over to be the Ace. If we count the letters ACE from the beginning we get to the fourth box. We put the Ace into this box. Next we count out another three cards TWO to reach the eighth box. We put the Two into this box. We illustrate this below

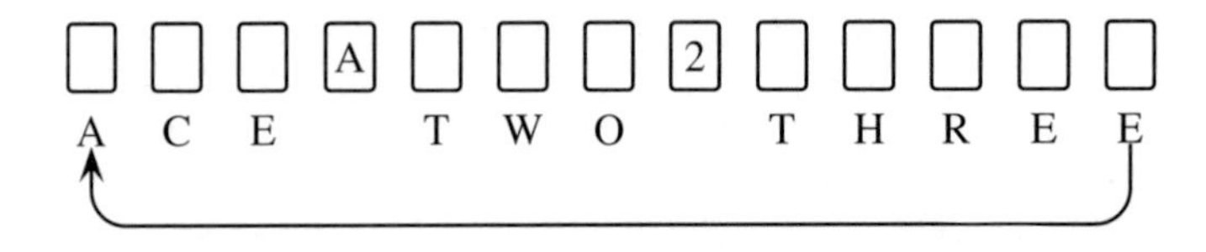

Now count out the letters THREE. By the time we have done this we have reached the end of the line of boxes. So where can we put the Three? Well, as we count out the cards we place them on the bottom of the pack. This means that when we get to the end of the boxes we simply start counting again from the beginning. Thus the Three is put into the *first* box. Now count out the letters FOUR. After counting out FO we get to the box with the Ace in it. But in the trick, this card will have been placed onto the table in front of us and is no longer in the pack. So we simply jump over it and count the two letters UR after it. The Four then goes into the seventh box.

Exercise. *Carry on with this process until you have placed each card into its own box. You should then end up with the cards in the order given above.*

Having seen how this trick works, you can now make up some variations of your own. For example you could make a set of 13 cards with objects of your choice on them, favourite pop stars for example, and work out the positions of the card so that each pop star knows there own name. The variations are endless.

The idea of having a set of objects which all fit into their own box is often called the pigeon-hole principle. Although it is a simple idea we have seen how it leads to a very nice trick.

Trick four: Mind reading

This is a marvellous trick which, if delivered with style and panache, is extremely mysterious. It uses some theorems in the branch of mathematics called number theory, to allow you to apparently read the minds of the audience. To do this trick the audience will need a pencil and paper and should also be able to do a straightforward subtraction.

What is the trick?

You ask several volunteers in the audience to choose a random five digit number in which not all of the digits are the same. Call this number x. Say to them that in order to make this number even harder to guess you want them to scramble the digits. In other words they should write down another five-digit number with the same digits as the original but in a different order. Call this number y. For example, if the original number is $x = 52441$, then the new number could be $y = 14254$. Now say that to make things even more random you want them to subtract the smaller of their two numbers from the larger to give a third number z. In this case we have

$$z = 52441 - 14254 = 38187.$$

They have now produced a number z which seems as random as possible. You will now use your great mental powers to try to find out z.

Ask each of your volunteers to choose *one* of the digits of z *which should not be the digit zero*, ask them to tell you the remaining digits, which you write down. For the example above they may choose the digit 8 and then tell you the remaining digits 3, 1, 8, 7. Now ask them to concentrate and to think of the hidden digit. After a moment of deep concentration you write down this mystery digit (hopefully to great applause). Continue to play the same trick with each of the other volunteers. The audience response will get better and better as you correctly guess more and more of the mystery digits.

How does the trick work?

Whilst they are 'thinking' of the mystery digit you add up the digits that they have given to you. Let's call this total t. Now find the nearest multiple of 9 to t which is greater than t. The mystery digit is then the difference between this and t. In our example, adding up the given digits we have $t = 19$. The nearest multiple of 9 which is greater than t is 27. Now, $27 - t = 27 - 19 = 8$, and we deduce that the mystery digit is 8. Even with practice, doing this calculation in your head takes a little time, and it is worth giving yourself a bit of extra time to check your calculation by saying things like 'you're not thinking hard enough, think harder' or such like to your willing volunteer.

Warning: Sometimes this trick does not work, the reason being that your volunteer has done their subtraction incorrectly. If you say what their digit is and they disagree, tell them that your great mental powers can see that they have made a mistake in their subtraction and ask them to check it! You will always be right!

Why does this trick work?

This trick gives us a small introduction to the wonderful branch of mathematics called *number theory* which is the study of the properties of the integers. The mathematical proof of why this trick works is rather more difficult than the mathematics behind the other tricks, so don't be too put off if you don't understand it at first. Rest assured, the trick will work every time even if you don't understand fully the reason behind it.

However, number theory is a branch of mathematics with many surprises, and locked within it are some of the greatest mysteries and unanswered problems of modern mathematics. (For example, it is still not known if every even number apart from 2 can be expressed as the sum of two prime numbers.)

Here is why the mind reading trick works.

Suppose that a number when expressed in base 10 has the digits $abcd \dots e$, for example the number 1234567. The number that we get when we add up all of the digits of this number is called the *digital root*. So the digital root is the number $a + b + c + d + \cdots + e$ and for our example this number is $1+2+3+4+5+6+7 = 28$. The digital root of a number has a lot of interesting properties and we will exploit some of them in this trick. The first property we use is that

> *when you take any number and subtract its digital root you always end up with a multiple of 9.*

For example, $1234567 - 28 = 1234539 = 9 \times 137171$. We will prove this result in two stages. Firstly, consider any *power of 10* such as 10, 100, 1000, etc. If we subtract 1 from any of these numbers we get a number such as 9, 99, 999. It is obvious that any of these numbers is divisible by 9. If you look at the pattern you will see that $10 - 1 = 9 \times 1$, $100 - 1 = 9 \times 11$, $1000 - 1 = 9 \times 111$, etc.

More generally, suppose that n is any *natural number*, for example $n = 1, 2, 3, 4$, etc. Then 10^n is a power of 10. Now, look at the following multiplication identities:

$$100 - 1 = 10^2 - 1 = (10 - 1) \times (10 + 1)$$

and

$$1000 - 1 = 10^3 - 1 = (10 - 1) \times (10^2 + 10 + 1).$$

You can check these by simply multiplying things out. All of this can be generalized to the following very useful identity

$$10^n - 1 = (10 - 1) \times (10^{n-1} + 10^{n-2} + 10^{n-3} + \cdots + 10 + 1).$$

Exercise. *Check this identity by multiplying everything out.*

Exercise. *Show further that the same identity also works if you substitute any number for 10.*

Now, $10 - 1 = 9$, and what we have shown is that

$$10^n = 1 + \text{a multiple of } 9.$$

Now, to say that a number x has the decimal digits $abcd \dots e$ is just a short-hand way of writing down the number as

$$x = a \times 10^n + b \times 10^{n-1} + c \times 10^{n-2} + d \times 10^{n-3} + \cdots + e$$

where $n + 1$ is the total number of digits. We will use this representation of x again in Chapter 7. For example, the number 1234567 has the digits 1, 2, 3, 4, 5, 6, and 7 and we express it as

$$1234567 = 1 \times 10^6 + 2 \times 10^5 + 3 \times 10^4 + 4 \times 10^3 + 5 \times 10^2 + 6 \times 10 + 7.$$

Now, it follows from the above identity that we can always write the number 10^n as

$$10^n = 9 \times k_n + 1.$$

Here k_n is the number with n digits all of which are the digit 1. For example $1000 = 10^3$ and $1000 = 9 \times 111 + 1$ so that $k_3 = 111$. Similarly $k_1 = 1, k_6 = 111111$ and we will let $k_0 = 0$. Using this expression we can then write the number x as

$$\begin{aligned} x &= a \times (9 \times k_n + 1) + b \times (9 \times k_{n-1} + 1) + c \times (9 \times k_{n-2} + 1) \\ &\quad + d \times (9 \times k_{n-3} + 1) + \cdots + e \end{aligned}$$

or simply

$$x = 9 \times (ak_n + bk_{n-1} + \cdots) + a + b + c + d + \cdots + e.$$

In other words, x is a multiple of 9 plus the digital root of x, which is what we wanted to prove.

In our trick we took a number x and then scrambled its digits to give another number y. Now the number y has *exactly the same digits* as the number x and therefore *has exactly the same digital root*. This simple observation is the key behind our trick.

Suppose that the digital root of x is the number d. We know that x is a multiple of $9 + d$, so that there is a whole number p with $x = 9 \times p + d$. In exactly the same way there is a whole number q with $y = 9 \times q + d$. Now, suppose that x is bigger than y and we then subtract y from x to give the mystery number z. It follows from the above that $z = 9 \times p + d - (9 \times q + d) = 9 \times (p - q)$. In fact we have reached the following remarkable conclusion:

The number z is a multiple of 9.

Exercise. *Try this out on any number x – it doesn't matter how many digits x has.*

Having shown that the mystery number z is a multiple of 9 we can now finish off explaining why the trick works. To do this we need to use the following result

> *The digital root of any multiple of 9 is also a multiple of 9.*

Exercise. *Show that if a number is divisible by 9 then the digital root of its digital root is also divisible by 9.*

You have known this result almost as long as you have known your times tables. If you take the first 10 multiples of 9

9 18 27 36 45 54 63 72 81 90

then it is easy to see that the digits of each of these numbers add up to 9. This is always a useful way of checking our 9 times table. The result above generalizes this.

It is not difficult to see why the result is true. Suppose that our number is z and that it has a digital root d. Then as z is a multiple of 9 there must be a number p so that

$$z = 9 \times p.$$

We also know from our earlier work that $z - d$ is always a multiple of 9, so that there is a number q for which

$$z - d = 9 \times q.$$

Putting these together we have

$$d = z - (z - d) = 9 \times p - 9 \times q = 9 \times (p - q),$$

so that $d = 9 \times (p - q)$ is a multiple of 9.

Note how we have sneaked in a use of the subtraction formula that we used in the clock tapping trick.

We have now just about finished explaining why the trick works. We have shown that the number z that we get when subtracting two numbers with the same digits in a different order is a multiple of 9. We have now shown that the digital root of this number is also a multiple of 9. Thus to find the hidden digit of z we start with the certain knowledge that the sum of the digits of z must be a multiple of 9. Thus if we are given all but one of the digits of z we add up the digits we have been given and then make the sum up to the nearest multiple of 9 and *hey presto* we have worked out the hidden digit. And if that doesn't bring the house down then nothing will!

4.4 Conclusions

We hope that you will enjoy performing some of these magic tricks in front of a live audience. At the end of your performance, no one will ever again think that maths can be dull. The beauty of mathematical magic, is that almost every branch of mathematics has surprising and mysterious results which can be turned into magic tricks. For example the mathematical theory of topology which we said a little about in chapter one leads to a whole load of wonderful tricks involving handkerchiefs! We suggest that you have a read of the books in the list at the end of this chapter, or better still make up some tricks of your own.

4.5 Exercises

First session

1. Do you have a favourite magic trick? Can you think of any magic tricks that you know of which might involve mathematics?
2. Practice the clock tapping trick.
3. As a variation on the clock trick, ask someone to count to 25 instead of 20. If you start the anticlockwise tapping from 12, how many random taps should you have before you do this. Alternatively, if you have eight random taps, where should you start counting backwards from?
4. Although it seems very different, this trick works in a very similar manner to the clock trick. Ask someone to take their age (don't tell you what it is) and then add it to today's year. Then ask them to subtract off the year in which they were born and to tell you the answer. Can you work out how to tell their age from this answer (try it out on an example to get the idea).
5. Ask someone with a watch to choose any number x on it between 1 and 6, but not to tell you what this number is. Ask someone else to tell this person (secretly) a number y between 1 and 20. Now ask the first person to count *anticlockwise* y taps from x and to remember the number p that they get. Then ask them to count y taps *clockwise* from x and to remember the new number q. Ask them to add p and q and tell you the answer z.

   Try this out a few times. Can you guess how to work out x from the number z (ask if you get stuck). See if you can work out why your answer is true.
6. Simplify the following expressions involving subtraction:

$$(a)\ y + x - (y - x) \quad (b)\ y - x + (y + x) \quad (c)\ (y - x) \times (y + x).$$

Second session

1. Practice the 21-card trick. Can you think of any variations to make the trick appear more dramatic?

2. Suppose that you use three piles of nine cards (giving 27 cards in all). Show that if a card is at a position x in the pile then the effect of dealing out the cards, putting the pile with the chosen card in the middle and collecting the piles (just as in the 21-card trick) is to move the card to the position

$$x \to 9 + \lceil x/3 \rceil.$$

Show (its probably good to try a few practice runs) that you can find the card using exactly the same methods which worked for 21 cards except that when you deal out the cards at the end, the mystery card is in a different position. What position is this?

Check that the trick also works if you use 15 cards dealt into three piles of five cards each.

3. Now try three piles of 10 cards each (making 30 cards). Try the trick in this case and see what happens to the position of a card at x. Will the trick ever work (does doing more than three deals help here)? If you think that it won't work, can you think of a way of changing the trick to make it work.
4. Are either of the following true?

$$(a)\ \lceil x \rceil + \lceil y \rceil = \lceil x + y \rceil, \quad (b)\ \lceil x \rceil \times \lceil y \rceil = \lceil x \times y \rceil.$$

If you think they are true, prove it. If you think they are false find some counterexamples.

5. Practice the 'cards that know how to spell' trick.
6. Devise some variations on this trick. For example how might you arrange the cards so that each time you count out the name of someone you know (say in your family or in a football team) before dealing out the card?
7. Here is an exercise in the use of the pigeon-hole principle. Suppose that there are n people at your magic show. Each person knows between 0 and $n - 1$ of the people (other than themselves) at the show. Let's assume that if person A knows person B then person B knows person A. Prove that there are *at least two* people at the show that know exactly the same number of other people.
8. Try out the mind reading trick on your friends.
9. Here is another trick which uses the digital root of a number. Take a pack of cards and put a *joker* as the nineth card in the pack counting down from the top. Ask someone in the audience to tell you any number between 10 and 19. Let this number be x. Count out x cards from the top of the pack, placing them face down in front of you, one on top of the other to form a new pile. Now ask your volunteer to add up the digits in their mystery number x to give the digital root d. Pick up the new pile and count d cards from the top. Now say to your volunteer that they could have chosen any number but they chose to be the joker. Turn over the next card from the pile in your hand. It is the joker.

Practise this trick and see if you can work out why it works.

Field trips and projects

Either visit a magic show or run one for yourself.

4.6 Further problems

One surprising aspect of mathematical logic is that it often seems to contradict our own intuitive reasoning, or notions of common sense. This can be exploited in a 'magical' way by seeing how much the views of an audience are in contradiction to the actual facts. In this set of further problems we will look at some examples of mathematics surprising us by contradicting what we think is going to happen.

An excellent way to do this is to look at the way that people perceive their chances of winning are when they are playing some sort of game. A good game to look at is a lottery, as most people take part in these. In the UK the National Lottery has been running since November 1994. It is an entirely fair process which runs on meticulously logical lines. In particular, anyone entering it has *exactly* the same chance of winning with one entry as anyone else. However, the public perception of the sort of results that you get from playing it are very different from the logical realities.

1. How likely are you to win the lottery?
The UK National Lottery works as follows. A rotating barrel is filled with 49 balls and set in motion. Six balls are then drawn out of the barrel. Sometime in the previous week you will have filled in your ticket, indicating six numbers of your choice. You win the jackpot if your six numbers and those of the machine agree. As the process of choosing the six balls is entirely fair and random, all combinations of balls are equally likely. Now, what are your chances of winning the jackpot? This is a calculation in probability. The chance that the first ball out of the barrel is one of the balls on your list is $6/49$. There are now 48 balls in the barrel and you have five numbers left on your sheet. The chance that the second ball is one of these is $5/48$. There are now 47 balls left and four numbers on your sheet. You will be on the roll with a chance of $4/47$ and at this point (in the UK) you have already won £10. But do we stop there? Oh no – we go on to the bitter end. The next ball is ours with a chance of $3/46$, the next with a chance of $2/45$, and finally (you can feel the excitement) the last ball is ours with a chance of $1/44$. When combining chances you multiply the probabilities to give an overall chance of hitting the jackpot of

$$\frac{6}{49}\frac{5}{48}\frac{4}{47}\frac{3}{46}\frac{2}{45}\frac{1}{44} = \frac{1}{13\,983\,816}.$$

Now, the problem with this calculation is that not many of us appreciate what a chance of $1/13\,983\,816$ really means. This is where our intuition and logic differ to a surprising degree. To make this plain we pose the following question

How do your chances of winning the lottery jackpot compare with your chance of living to see the result?

In fact – pretty badly. Let's see why. The main cause of death amongst the school age group is a car accident. Now about 50 people are tragically killed per week in a car accident in the UK, meaning that out of a population of 56 000 000 your chances of being killed in any given week in a road accident are about $1/1000\,000$. Let's compare this with your chance of winning the jackpot. We see an uncomfortable truth

You are about 14 times more likely to be killed in any given week than you are to win the jackpot on one ticket.

Or, to put it another way, you are only going to have an equal chance of living and of winning the jackpot if you purchase your coupon on Saturday afternoon!

2. Birthday coincidences

Let's have another look at where logic and our intuition are surprisingly at odds. Many people base their choice of lottery numbers on their birthdays. Suppose that you are doing a magic show with, say, 30 people in the show. What are the chances of two of them having the same birthday (and perhaps by implication the same choice of lottery numbers)? If you did not think very hard you might guess that it would be rather unlikely; maybe you would reason that as there are 30 people and there are 365 days in the year then the chance of two having the same birthday would be something like $30/365 \approx 1/12$. However, this reasoning is wrong and the actual answer is very surprising. For this size of group, there is a much better than evens chance that two of them will have exactly the same birthday. You can use this as a trick to dazzle the audience of your magic show with your ability to 'predict' when they will have their birthdays. The calculation of this results goes as follows, and in fact this calculation gives us a good way of doing the trick in a magic show.

Line all of the people in the show in a single line and get each of them to call out their birthdays as you go down the line. Write these birthdays down. The trick is to see whether any of these two birthdays are the same.

Now, this sort of magic trick does not always work, so lets see how likely it is that the people all have *different* birthdays and we will look a failure in front of the audience! The first person in this line will have a birthday at some point in the 365 days of the year. We will assume that any day is as likely as any other for someone to have a birthday on. Now take the second person. If they are to have a different birthday from the first person then there are only 364 days on which their birthday can occur. Thus the probability that their birthday differs from the first persons is $364/365$. Looking at the third person, for them to have a different birthday from the first two, their birthday must fall on one of the 363 remaining days. This will occur with probability $363/365$. We can continue this process. If our line has 30 people in it then the chance that they will all have different birthdays is

$$\frac{364}{365}\frac{363}{365}\frac{362}{365}\cdots\frac{337}{365}\frac{336}{365}$$

which is

$$0.2927\ldots$$

In other words, the probability that two of them will have the same birthday is $1 - 0.2927\ldots$ which is a staggering

$$0.7063\ldots$$

or better odds than 7 to 10. Certainly more worth taking a gamble on than the lottery. Of course the trick may not work, but with 30 people in the show it is very likely that it will.

We can make a variation on this trick by asking how many people we need in the show to have a better than evens chance that two of them will have the same birthday. If there are n people in the show, then the chance that they all have different birthdays is

$$\frac{364}{365}\frac{363}{365}\frac{362}{365}\cdots\frac{(367-n)}{365}\frac{(366-n)}{365}.$$

Exercise. *Check this result.*

The question is then, how large does n have to be for the chance of *not* having two birthdays the same to be just less than $1/2$? If you try this formula for various values of n you will find that when $n = 23$ then the chance of two people not having the same birthday is 0.4927. So, having 23 people in the show gives you a slightly better than evens chance that two will have the same birthday.

Exercise. *How many people do you need to* guarantee *two will share a birthday? (Hint: look back at Trick Three)*

3. *How do people choose lottery numbers?*

Having looked at the statistics of birthdays, we will now have a look at the statistics of the number of jackpot winners of the lottery for every draw. As we said above, the chance of winning the jackpot is $1/13\,98\,3816$. On average in the UK there are about 30 000 000 entries for each draw (more if there is a roll over). Thus, on average we should expect to see $30\,000\,000/13\,983\,816$ or about two winners per week. Does this agree with the facts? Well here are some statistics. In the first 30 weeks of the operation of the lottery in the UK in 1995:

- there was one week in which 133 people won the jackpot
- there were six weeks in which no-one won the jackpot
- eight weeks when one person one the jackpot
- only four weeks when two people won the jackpot.

These statistics differ markedly from what would be expected from an average number of two winners per week. As the numbers from the lottery are completely fair and random we are forced to the conclusion that

The way that people choose their lottery tickets is not random.

This again shows the difference between our perception of how a random sequence should behave and the way that it actually does behave. Machines are very good at choosing random numbers, but people seem to be very bad at it. In other words people often choose the same numbers as someone else – somehow their intuition about what a truly random number is, is letting them down. Here is an example of the way that people tend to choose their lottery draws. Consider a lottery ticket. This is divided into five columns. An almost instinctive way to choose your numbers is to place one in each column and then place the sixth reasonably distant from the ones that you have already chosen. This gives a nice even spread of numbers and the feeling that this is a nice random sequence. In fact it is anything but random. To show this we make the following definition.

A sequence of six numbers in ascending order contains a *pair* if two of the numbers differ by at most one. For example the sequence 2 10 18 19 25 39 contains the pair (18, 19) whereas the sequence 2 10 18 20 25 39 does not.

How likely is the lottery machine to pick a sequence with a pair. Most people think not very. Check by taking a straw poll of your friends to see what people think. A survey of a 'typical' Bath–Bristol Masterclass produced answers ranging from 1 in 100 to 1 in 10. No one came up with the surprising answer that the chance of getting a pair from the machine is *almost a half*. That is, on an average lottery draw you are about as likely to see a pair as not. You can check this very easily by recording the next ten draws from the lottery. In other words, of all possible combinations of numbers coming out of the lottery machine, nearly half of them contain a pair. Now, the method of choosing numbers by putting them into a nice spread tends to mean that you do not choose combinations with a pair. Thus, if you look at all the choices of sequences made by the public for any one lottery draw then the greater majority will not contain a pair. If the actual number coming out of the machine does not contain a pair then there may be several winners and if it does contain a pair then there may be no winners at all. A more extreme example of this occurs when the draw contains a *triple*, that is three numbers in sequence. Again, you may think that this is very unlikely to happen. However, the chance of a triple happening is about 1/20, so over the year (with 104 draws) you are likely to see it happen about five times – which is quite a lot if you think about it. You actually improve your expected winnings if you choose a pair or a triple. Not because you are any more likely to win. Indeed, and this is most important, all combinations of numbers are equally likely. However, in the unlikely event that you do win, it is quite likely that you won't have to share your winnings with anyone else.

4.7 Answers

First session

3. If your mystery number is x then counting to 25 means that the clock will be tapped $26 - x$ times. When you tap the clock anticlockwise from 12 to reach the number x you make $13 - x$ taps. Therefore the number of random taps is $26 - x - (13 - x) = 13$. If, instead, you have eight random taps, then you need to make five more taps on the clock, so you should start counting from 5.
4. The nice thing about this trick is that not only can you work out someone's age, you can also tell roughly what part of the year they were born in. Suppose that the person's age is x and today's year is y. The first number that you work out is $x + y$. Now, suppose that z is the year in which they were born. In this year they will have a birthday. If they have already had their birthday then $y = z + x$. If they have not had their birthday then $y = z + x + 1$. The number calculated is $x + y - z$. If they have had their birthday then this is $x + (z + x) - z = 2x$. If they have not had their birthday it is $x + (z + x + 1) - z = 2x + 1$. So, if the total that they give you is *even*, divide it by two to give their age x and say 'you have already had your birthday this year haven't you'. If the total they give you is *odd*, then subtract one and then divide by two to give their age x and say 'you have a birthday coming up this year'. You could combine this trick with the birthday coincidences trick described in the further problems.
5. Counting 20 anticlockwise on the clock starting at x is like subtracting 20 from x. The difference is that we are subtracting 20 modulo 12. In the chapter on codes and ciphers we will go into this in detail, however, for the purposes of this answer, the number that we get on the clock face is $x - 20 \bmod(12)$, or in other words you add a multiple of 12 to $20 - x$ until you get a number between 1 and 12. Similarly, counting clockwise 20 from x is the same as adding 20 to x modulo 12. Thus

$$p = x - 20 \bmod(12) \quad \text{and} \quad q = x + 20 \bmod(12).$$

Combining these two results we have that

$$p + q = 2x \quad \text{plus a multiple of 12.}$$

If $p + q$ is less than or equal to 12 then divide it by two to give x. On the other hand, if $p + q$ is greater than 12, then subtract 12 and then divide by 2. Let's see how this works with two examples. If $x = 2$ then $p = 6$ and $q = 10$. Now, $p + q = 16$ which is bigger than 12. Subtracting 12 we get 4 and dividing by two we get 2, which is x. On the other hand if $x = 5$ then $p = 9$ and $q = 1$. In this case $p + q = 10$ and half of this is 5 which is x.

Now, in the question we insisted that x should lie between 1 and 6. The reason for doing this is that if we do not make this restriction then we can't find x uniquely. For example, if $x = 8$ then $p = 12$ and $q = 4$ so that $p + q = 16$. This is exactly the same result which we obtained earlier when $x = 2$. The reason for this is that

$8 - 2 = 6$ and $2 \times 6 = 12$ so that both 2 and 8 satisfy the same equation $2x = 4$ mod(12). This seems a little strange and we will have another look at this property of modular multiplication in Chapter 6 on codes and ciphers.

6. (a) $y+x-(y-x) = 2x$, (b) $y-x+(y+x) = 2y$, (c) $(y-x)\times(y+x) = y^2-x^2$.

Second session

3. Suppose that we have a pack of 30 cards. If the card is at the position x, then the formula for the new position y of the card after one deal is given by

$$y = 10 + \lceil x/3 \rceil.$$

If you look at this map for yourselves you will find that after three deals (as before) if the card is in position 1 up to 15 then it gets to position 15 and stays there. On the other hand if it is in position 16 to 30 then it gets to position 16 and stays there. In other words the map has *two* fixed points at 15 and at 16. The simplest way of working out which it is, is to then deal again into the three columns. If the card selected is in the *first* column it must be card number 16, on the other hand if it is in the *last column* then it must be card number 15. Note the card can never be in the middle column on the last deal.

4. (a) Untrue – take $x = 2.6$ and $y = 2.6$.
   (b) Untrue – take $x = 2$ and $y = 2.6$.

7. There are two cases to look at. Suppose that there is at least one person that knows no-one else. Then there cannot be anyone who knows them, and so the most people that someone else can know is $n - 2$ people. So, each person can know between 0 and $n - 2$ other people. This means that there are $n - 1$ different alternatives for the number of people that one person can know. Imagine now that the people at the show are boxes and in each box we put a number which is how many people that they know. There are n boxes and there are $n - 1$ different alternatives for the number of people that each person knows. Because of this there is no way that each of the n boxes can contain a different number taken from the set 0 up to $n - 2$. In other words at least two of the boxes must have the same number in it, which means that at least two people at the show must know the same number of other people.

   The other case is when there is no person who knows no-one else. In this case the different number of people that they know must be a number between 1 and $n - 1$ which again gives us $n - 1$ choices, so we are back to the situation we looked at above.

9. This trick is nothing other than a disguised version of the result that subtracting the digital root of a number from the original number gives a number which is a multiple of 9. In particular, if you take any number between 10 and 19, then that number minus its digital root is simply 9. The process of dealing out the cards in the order described is a way of doing this subtraction automatically. You can use this trick as a way of forcing your volunteer to choose a card which you have decided in advance.

4.8 Mathematical notes

Formulae for π

There are many formulae to calculate π. The ones given in this chapter converge to π very slowly, but we can do much better. One that converges very rapidly indeed is the following formula discovered by Ramanujan,

$$\frac{426\,880\sqrt{10\,005}}{\pi} = \sum_{n=0}^{\infty}(-1)^n \frac{(6n)!(k_1 + nk_2)}{(n!)^3(3n)!(8k_3k_4)},$$

where $k_1 = 13\,591\,409$, $k_2 = 545\,140\,134$, $k_3 = 100\,100\,025$ and $k_4 = 327\,843\,840$. Remember that $k! = k \times (k-1) \times \cdots \times 2 \times 1$.

Just taking the first term of the sum (that is taking only the $n = 0$ part) gives

$$\pi = \frac{53\,360\sqrt{640\,320}}{13\,591\,409} \approx 3.141\,592\,653\,589$$

and adding the first two terms gives π correct to 40 decimal places!

National Lottery

The *National Lottery* provides a rich source of many problems in probability, and the conclusions of these will be of immediate interest to many people. Interesting long-term projects can include a careful look at the numbers that come up in the lottery every week. The probability of an individual number (say 39) coming up in a single draw is $p = 6/49$ or about 1/8. Over a period of some weeks, in which there are n draws then the expected number of times that a particular number will come up is $\mu = 6n/49$. Now, if n is large, as p is relatively small, the number of occurrences of the particular number will form, approximately, what is known as a Poisson distribution. In other words, the probability that the particular number will come up r times in the n weeks is approximately given by

$$\text{Prob}(r) = e^{-\mu}\frac{\mu^r}{r!}.$$

For example if $n = 47$, then the mean number of times that a ball will come up is 5.755. The probability that a certain number will not come up at all in the n draws is

$$e^{-5.755} = 0.00316\ldots,$$

and the probability that a number will only come up only once is

$$5.755e^{-5.755} = 0.01822\ldots.$$

The expected number of balls which will not have turned up in this 47-week period is $0.00316 \times 49 = 0.154$ and the expected number which will only come up once is $0.01822 \times 49 = 0.896$ which, significantly, is quite close to one. In the first 47 weeks of the UK National lottery the number 39 only appeared once. This seemed very

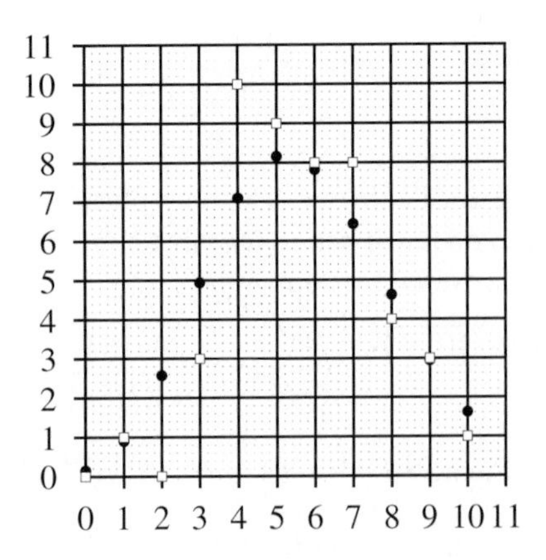

Figure 4.2: Comparison of E_r (dots) and O_r (squares) for the National Lottery.

surprising and various explanations were given in the press for why this happened. Most discussed the significance of the digits on the ball and how this might affect the way in which it came out of the machine. But in fact we can see from the above that 'on average' in a completely fair lottery we would expect one ball to only come up once in this number of draws – so there is no evidence at all that there is anything exceptional about the number 39. Here for comparison are the actual statistics for these 47 weeks of the observed number O_r of balls that come up r times compared with the expected number of balls E_r that would come up in a Poisson distribution where

$$E_r = 49\, e^{-\mu} \mu^r / r! \quad \text{and} \quad \mu = 47 \times (6/49).$$

r	0	1	2	3	4	5	6	7	8	9	10
O_r	0	1	0	3	10	9	8	8	4	3	1
E_r	0.15	0.90	2.57	4.94	7.09	8.16	7.82	6.43	4.62	2.95	1.62

You can see in Figure 4.2 that there is a very reasonable agreement between these two figures, exactly as you would expect from a fair lottery. Now, try collecting statistics of the lottery draws for yourselves over a large number of weeks to see if they agree with the results expected from the Poisson distribution.

Geometry

We have so far avoided geometry, but this seemingly completely straightforward branch of mathematics has its own magical surprises. Here is one of the best which will certainly catch out the unwary. Take a circle and place two dots round the outside. Join up the dots with a line. This divides the circle into two regions. So far so good. Now place three dots. Join a line from each dot to every other dot (you should have three lines) and count how many regions that you have. Repeat this for four dots and

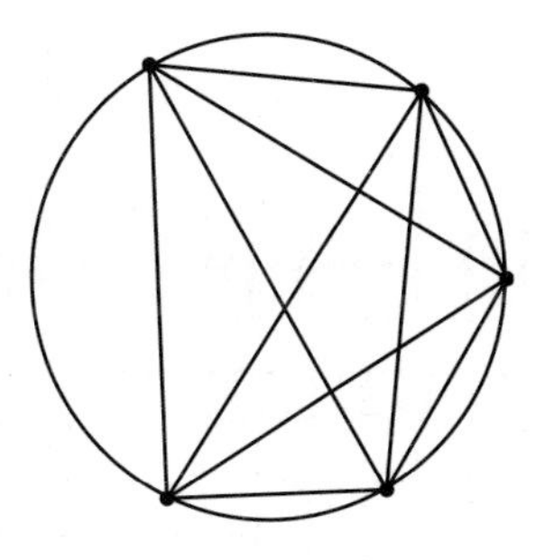

Figure 4.3: Five dots gives 16 regions.

for five dots. Each time count the regions. An example with five dots is shown in Figure 4.3.

You will get the sequence 1, 2, 4, 8, 16 for the number of regions. Now consider six dots. Using your results it would seem only reasonable that there should be 32 regions. In fact there are 31. Somewhere we seem to have lost a region. For seven dots there are 57 regions instead of the 64 we might have expected. Obtaining the general formula is a tricky application of combinatorics. In general, if d is the number of dots, then the number of regions r is given by

$$r = C_4^d + C_2^d + 1 = \frac{1}{24}\left(d^4 - 6d^3 + 23d^2 - 18d + 24\right).$$

We leave it as an exercise for you to work out why this formula works.

4.9 References

There are many books on magic tricks and if you look through them you will probably find that a lot of these tricks are actually based upon mathematical principles. As a challenge, try looking through some of these books to see if you can detect some of the underlying mathematical principles.

- Gardner, M. (1955). *Mathematics, Magic and Mystery.* Dover.
  This book is the best by far on the relationship between mathematics and magic. It has a very large number of tricks in it all based on mathematical principles ranging from number theory through card tricks to amazing things to do with a handkerchief. Martin Gardner has written many fantastic popular books on mathematics which you will find at end of several of the chapters of this book.
  Other books include:
- Blum, R., and Sinclair, J. (1991) *Mathemagic.* Sterling Publishing.
- Heath, R. V. (1967) *Mathemagic.* Dover.
- Blatner, D. (1998). *The Joy of Pi.* Penguin.
  This book explores in detail the history of π.

5
Castles: mathematics in defence and attack

5.1 Introduction

Usually in a workshop we aim to work with each other, to do things in teams, and to generally make friends. This chapter is an exception as we consider how people can be seriously nasty to one another and how a mathematician can help them to do it. However, although the class involves some practical experiments, no loss of blood will be involved – at least not on purpose. What we will do is to use mathematics to understand how and why castles were constructed the way they were, and how a bit of mathematics can help us both to design them to be easier to defend and also make them easier to attack. So you thought that castles were just to be studied in history lessons? Well read on!

The business of building fortifications to protect you and your family from the next person and their family is as old as human civilization. It is a sad fact that often the main mark that remains of a civilized society are the fortifications that they erect to defend themselves from another civilization! Indeed, it has been said that the only man-made object that can be seen from space is the Great Wall of China, which was built around 240 BC to defend the Chinese people. Thus, at first glance, an alien might only be able to tell that we were civilized by looking at one of our defences. When excavating an ancient city often the first thing that you look for is evidence of the fortifications that were used to defend it. One of the oldest known fortified cities is Jericho, and the fortifications around Jericho are described in the Old Testament in the sixth chapter of the book of Joshua. In this story, Joshua breached the walls by using trumpets and the help of God. In this chapter we will consider more conventional means of attack.

The word 'castle' comes from the Roman words *castellum* meaning a fortified camp and *castrum* meaning a fortified place. However, fortifications are a lot older than the Romans and some are still in use today. The classical design of the castle is usually associated with the fortifications built during the Medieval period. In this chapter we will ask one basic question, namely, what is the best way to build a fortification so that it is easy to defend? We will show that by using some mathematical ideas it is possible to come up with a solution to this problem. To solve it we will take a tour through geometry, the study of convex shapes, and the modern theory of fractals. The beautifully geometrical solution to this problem arrived at in the design of relatively modern fortifications, such as the Vaubain forts described in Section 5.4, gives a striking example of the role that mathematics has played in the lives of many people.

5.2 Early fortifications, circles, and the isoperimetric theorem

Some of the earliest fortifications were used to defend whole towns. These included the walls of Jericho and also the walls of Troy (which were built around 1500 BC). For ease of defence, many fortifications were built on top of hills and are called hill forts. Because of the size and indestructibility of these fortifications, they can often be seen today. In some cases hill forts are up to 3000 years old, and many were built during the Iron Age (approximately 700–55 BC). A lot of people lived inside hill forts, and they were also used to store grain, cattle, and other foodstuffs. Think back to those times. Inside the fort meant safety, but outside lay a cursed earth. In the UK there are at least 3000 splendid examples of hill forts. Here is a small selection of some of the best.

British Camp	Hereford
Caburn	E. Sussex
Hammerwood	E. Sussex
Old Sarum	Wiltshire
Figsbury Rings	Wiltshire
Yarnbury	Wiltshire
Windmill Hill	Wiltshire
Old Oswestry	Salop
Crickley Hill	Glocs.
Hod Hill	Dorset
Maiden Castle	Dorset
Dundry Hill	Bristol
Danebury	Hampshire
The Caterthuns	Angus
Walbury	Berkshire
Haddenham	Cambs.
Briar Hill	Northants.

One of the largest, best preserved, and most impressive of these fortifications is Maiden Castle in Dorset, which is currently open to the public. In AD 44 there was an epic battle at Maiden Castle when the invading Romans under the general Vespasian besieged the Iron Age tribe defending it. It was finally captured after a long siege and great loss of life. In the twentieth century the archaeologist Sir Mortimer Wheeler excavated Maiden Castle and found skeletons there which had evidence of having suffered a violent death, including one which has a spearhead in its spine and another which has an arrow head from one of the Roman siege ballistas impaled in its skull.

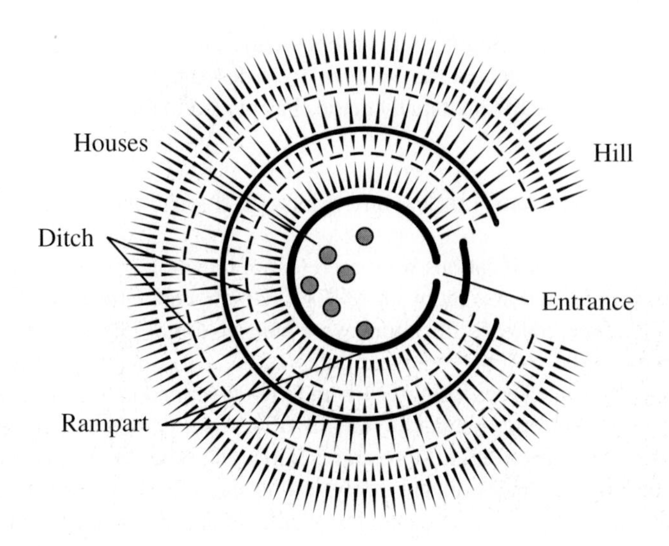

Figure 5.1: The typical layout of a hill fort.

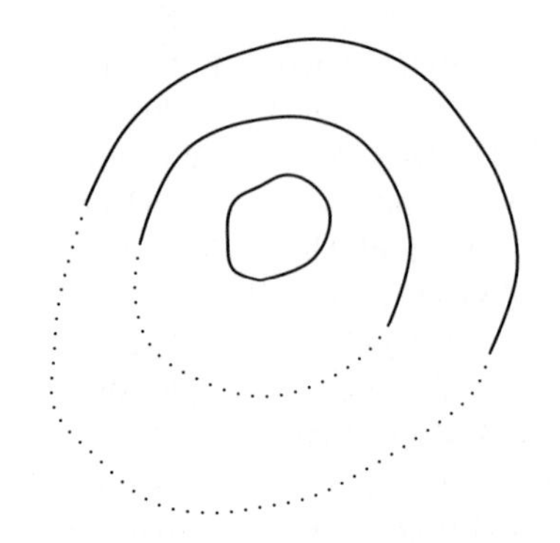

Figure 5.2: The ditches at Windmill Hill.

A typical schematic layout of a hill fort is shown in Figure 5.1. In Figure 5.2 we give the actual layout of the hill fort at Windmill Hill, Wilts., showing both the original and remaining ditches.

When building such a fort several ditches were constructed at the top of the hill in the form of a set of concentric rings. These gave a series of obstacles to be overcome by the attackers (who were of course coming up hill) and gave defence of the fort in depth. Between each ditch a huge rampart of earth and rock, excavated from the ditch, was thrown up. This rampart acted both as a barrier to the attackers and also a point from which the defenders could throw missiles. In the photographs in Figure 5.3 you

Figure 5.3: British Camp, Herefordshire, showing the ditches and ramparts.

can see some of the ditches and ramparts surrounding British Camp in Herefordshire. As an inner line of defence a large fence of sharpened posts was constructed. The forts also had elaborate entrances to slow down an attacker making for the easiest way in. Inside the ditches and fences you built your town and stored all your food. Building a fort on a hill had several advantages. The hill gave all-round visibility to the defender and the fort could be seen from a long way, giving a psychological advantage over the attacker. Also its position as a vantage point gave the defenders time to prepare for an attack. Climbing a hill was difficult for an attacker and it was easier for a defender to throw things down. Furthermore, hills were less swampy than the surrounding countryside. This was an advantage but it also meant that water supply was a problem, particularly when under siege. Other resources, such as food and the poles for the fences, had to be carried up the hill and they could not have been easy to build or maintain. It has been estimated that around 20 000 man days of effort were needed to build a typical hill fort. Similar methods of construction, namely concentric rings of ditches and ramparts, were also used in prehistoric sites of a more religious nature such as Avebury, Woodhenge, and Stonehenge.

The geometry of a hill fort

Although there is no clear historic record that the builders of hill forts attended courses in mathematics, it is certainly true that the design of a hill fort involves some mathematical principles. We will now have a look at three aspects of the design which demonstrate them, namely the design of the perimeter fence, the design of the entrance to the fort, and the location of the ditches and ramparts.

The design of the outer perimeter

If you look at the plan of Windmill Hill in Figure 5.2 you will see that its shape is very close to a series of concentric circles. By using a bit of mathematics we can work out several reasons why this is a good design.

You as the constructor of a hill fort are faced with the following problem. You have a limited number of posts at your disposal, and you want to make the perimeter fence as short as possible, for the simple reason that the longer the fence is the harder it is to build and the more defenders that you will need. However, you also want to get as many people as possible inside the fort. If this is your only consideration, what shape should you build the hill fort? This problem can be formulated mathematically. Suppose that the perimeter of the fort is a curve which is of a fixed length (dictated by the number of poles in the fence) or the number of defenders you have. To be a good defence this curve must be *closed*; this means that its ends must join up. Think of a piece of string with the ends tied together if you like. By changing the position of the string you can make many different shapes. Below we see several examples of closed curves.

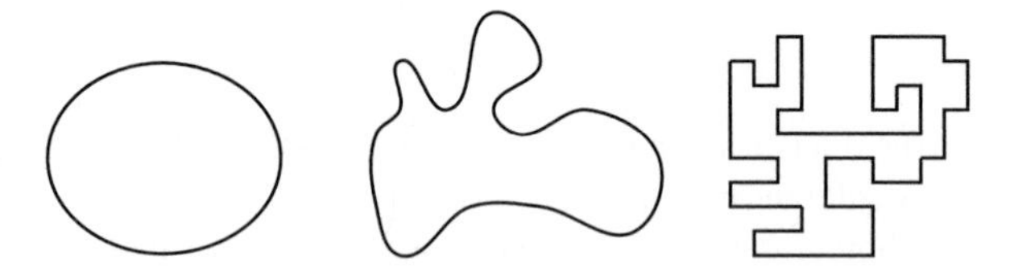

Now, look at the shape enclosed within the closed curve. This will have an area. The greater the area, the more houses that you can build inside, and the more defenders that you can enclose. So, now we ask our question – which of these shapes has the greatest area? So, we state our design problem as follows:

What closed curve of a given perimeter encloses the greatest area?

This question has a surprisingly easy answer which you may have already guessed from the design of Windmill Hill. If we can make any shape with a piece of string with a constant perimeter, then the shape which encloses the greatest area is simply

This result is so important that it has a special name. It is called the *isoperimetric theorem* which comes from the Greek meaning 'same perimeter'.

You can get an idea of why the shape with the largest area is a circle through a practical *demonstration*.

We need ten or twelve of you to hold hands (yes) and make up a closed curve. Imagine that you are the fence posts in the perimeter wall. Firstly arrange yourselves in a rectangle

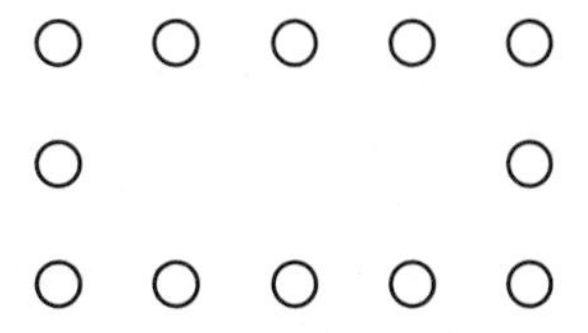

and see how many people you can pack inside the curve. Now, without letting go of each other's hands, open up the curve into a square. Now see how many people you can pack inside. Finally open out into a circle. You will find that you can get a surprisingly large number of people inside. Try for a record and see how many are possible. Will any other shape get as many people inside? This demonstration should help convince you that the shape with the greatest area is a circle, although it is not a proof.

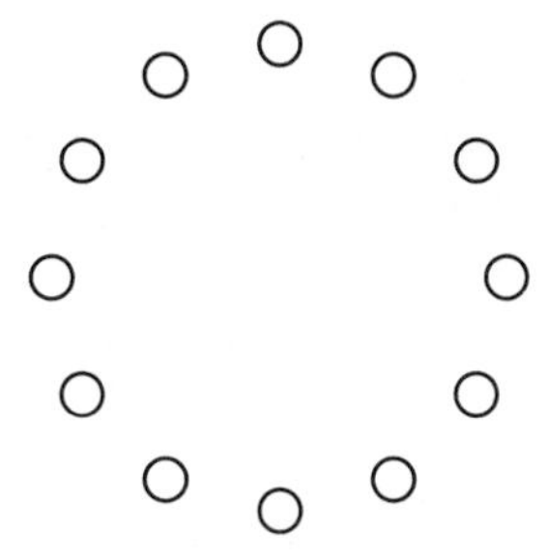

In the further problems we will prove this result – that is, show it is true by an utterly convincing and mathematically watertight argument. Although we will leave the details till then, the basic argument runs as follows.

Firstly, the shape with the greatest area has the property that if you take any two points inside it (these could be the defenders of the fort) then every point on the straight line joining them is also inside the fort. Shapes like this are called *convex*. In Figure 5.4 the first two shapes are convex and the third is not.

Exercise. *What features of a convex fort might make it easier to defend?*

Secondly, the shape with the greatest area is *symmetric*, so that it looks the same when it is reflected in a mirror lying on its centre line. Look at Figure 5.4. The first two shapes are symmetric and the third is not.

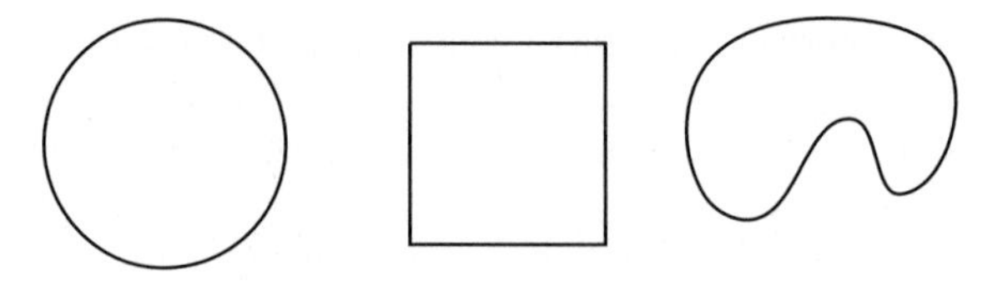

Figure 5.4: Some closed curves.

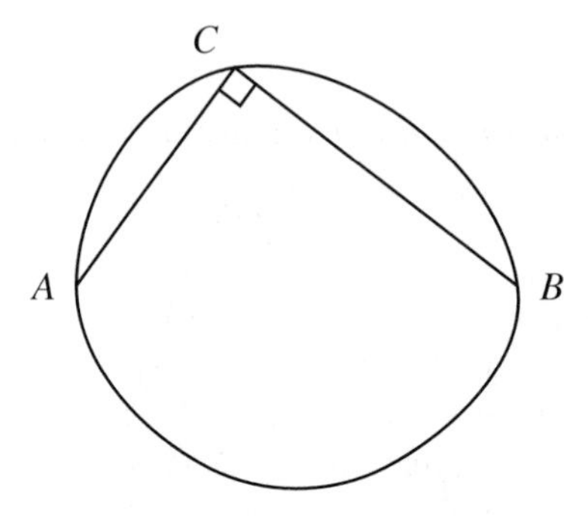

Figure 5.5: What shape always has a right angle at the point C?

Thirdly, if you take a centre-line AB through the shape and take *any* point C on the perimeter then the angle ACB is always 90°. This is shown in Figure 5.5. (If you've just asked 'why' then skip forward to the full explanation in the further problems.) We are now in a position to draw our shape which has the greatest area amongst all shapes of the same perimeter.

Start by drawing a line and mark the ends A and B. Now draw a line at an angle α to AB going through the point A and a second line at an angle $90^\circ - \alpha$ to AB going through the point B. Mark the point where they cross C. This is shown in Figure 5.6 (a) and this is a point on the perimeter the curve as the angle ACB is then automatically 90°. Repeat this construction for a range of values for the angle α between 0° and 90°. For example you could take $\alpha = 5^\circ, 10^\circ, 15^\circ, 20^\circ$, etc. Now do the same thing on the other side of the line AB. Figure 5.6 (b) shows the idea.

Try this out and look at the curve which passes through all the points marked C in the diagram. You will quickly convince yourselves that this curve is a circle. (We will show why in the further problems.)

Exercise. *Draw the shape that you get when the angle ACB is different from 90°, for example 45°. Can you show that this is also a circle with AB now being a chord rather than a diameter?*

The isoperimetric theorem demonstrates that a good shape for the outer defences is a circle. There are several other reasons for choosing a circle. One simple one is that

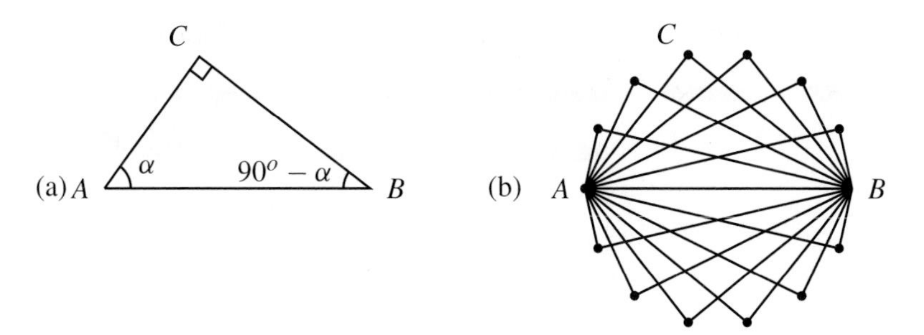

Figure 5.6: Constructing right angled triangles.

Figure 5.7: A hill fort hut.

as a circle is completely symmetric then every point on the perimeter is as strong as every other point. Thus it has *no weak points*. This is very important for a defender, as an attacker will always attack a fortification at its weakest point. In the next section we will see what happens to a castle keep when it has a weak point!

Circles are not always the best possible shapes. For example, the conditions at the top of the hill (particularly the amount of level ground) made a true circle impractical for very large hill forts such as Maiden Castle, and instead a more elliptical profile, following the high ground, was used.

The isoperimetric theorem applies not only to the design of the outer perimeter of the fort, but also to the design of the houses within the fort. When a fort is excavated, one thing that is quite easy to find is the shape of the houses. These are nearly always *circles*. A reconstruction of a hill fort house can be seen in Figure 5.7.

Why are they circles? We apply the same logic as before to see why. In times of shortage of building materials and lack of heating, the best design for a house is one

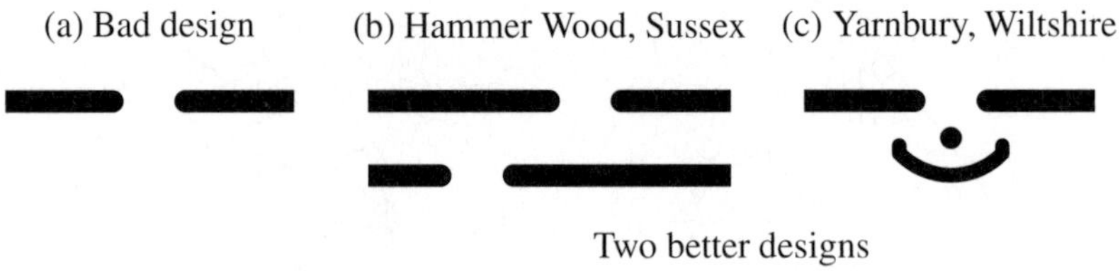

(a) Bad design (b) Hammer Wood, Sussex (c) Yarnbury, Wiltshire

Two better designs

Figure 5.8: The design of a hill fort entrance.

for which the outer wall is as short as possible. This is because the longer the wall is, the more materials are needed to build it and the more heat that is lost through it. However, we still want to get as much space inside as possible. And what shape does this best? – the circle!

The design of the entrance

Perhaps the most vulnerable point of a hill fort was its entrance. Obviously a gap in the ditches and ramparts is needed so that you, your family, and friends can get into the fort. However, if you can get in then so can an attacker. An entrance like the one illustrated in Figure 5.8 (a) would be difficult to defend as an attacker can enter easily without coming under fire from the defenders. A better entrance is one where the attackers are forced to spend a lot of time close to a defended rampart in order to enter the fort. However, you don't want the entrance to be too large as there is probably not too much space on top of the hill. Good examples of entrances are given in Figures 5.8 (b) and 5.8 (c) taken from different existing forts where we can clearly see that the attacker has no option but to run the gauntlet of spending a lot of time close to the ramparts if they want to enter, but at the same time are always close to the castle and thus in range of the defences.

In the exercises you can calculate how far an attacker would have to travel in order to get in to Maiden Castle which has a splendid entrance (from the point of view of a defender) which is illustrated in Figure 5.9.

The problem of designing an effective entrance for a hill fort can be stated as follows.

> Can you create as long a path as possible for the attacker to be forced down whilst they are close enough to the hill fort for you to attack them?

In fact this is really the question we considered in Chapter 1 when we looked at good ways of designing a labyrinth. The extraordinary mathematical fact is that in a given space close to the fort we can design an entrance in which the attacker can be forced to spend as long a time as we wish in range of the defences.

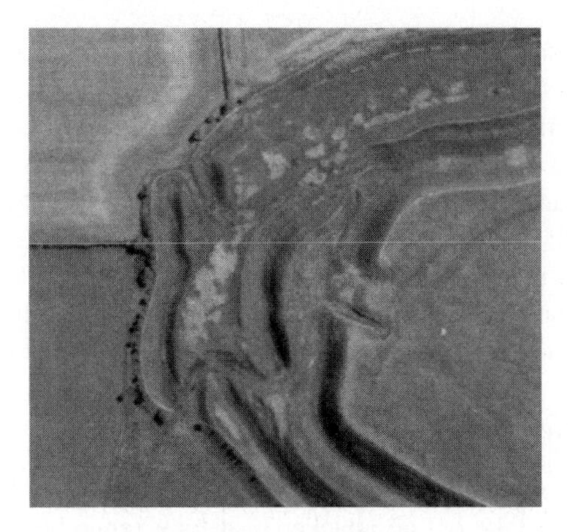

Figure 5.9: An aerial view of an entrance to Maiden Castle.

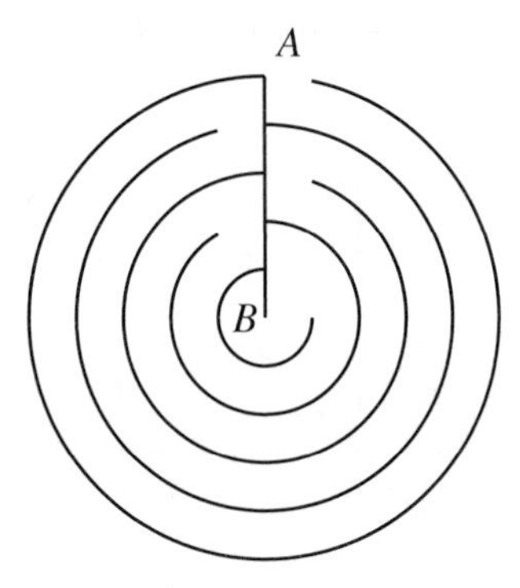

Figure 5.10: The walls of Jericho.

To find out why we will go back a few thousand years and look at the design of the walls of Jericho which were mentioned in the introduction. Tradition has it that they were a series of concentric circles with a final circular wall around the town. Each concentric ring had a gap in it and they were joined by a wall that not only provided a barrier but also provided access to the top of the wall for the defenders. Figure 5.10 shows possible picture of the walls.

If this figure seems familiar to you then have a look at the labyrinth drawn on the *Mappa Mundi* in Figure 1.6. To go from point A to point B when entering Jericho, the attacker has (in this figure) to go round four circles and is very vulnerable to attack at every point. If you want the attacker to go further, then just build a few more walls. (See the exercises.) Eventually we would have to stop because the gap between the walls would be too small for anyone to get through. However, mathematically we can take this process further – indeed there are wonderful curves called *fractals* which are infinitely long but which only occupy a finite amount of space. These would then be perfect entrances if attacked, although rather impractical if you needed to let your friends in! In the further problems we will look at a special fractal called the Koch snowflake which has an infinitely long perimeter but encloses a finite area.

A very long entrance in a small space is great from the point of defence but is of course a bit tedious if you want to use the entrance on a day-to-day basis. The entrance to Maiden Castle is a compromise between the optimal mathematical solution and the normal requirements of the inhabitants of the hill fort who might want to go out of the fort quite often and would not like to take an infinite time to do it.

Where do you put the ditches?

We can use some more mathematics to aid in the design of the ditches around a hill fort. The best way to take out an attacker is at long range before they have got to you and when they are at a comparative disadvantage. The most difficult point for an attacker is when they are at the bottom of the ditch and you have a clear view of them from the top of the rampart – see the illustration below of the fence and ditch at Rainsborough Fort, Herts.

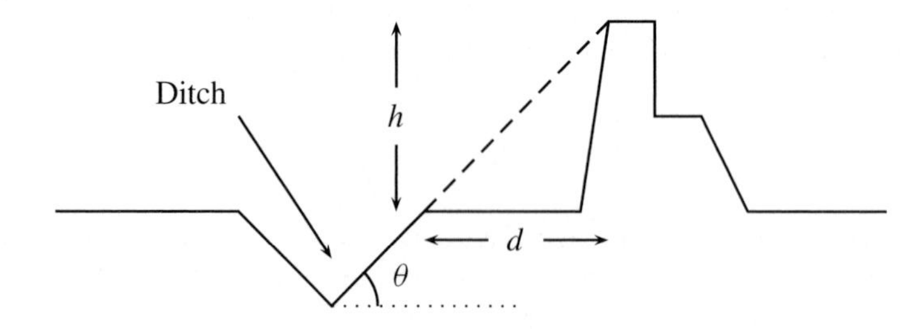

From the top of the rampart you can see into the ditch and can throw a rock at the attacker. We know that this was how the forts were defended, as the excavations of Maiden Castle turned up 20 000 pebbles at the entrance. A prehistoric ammunition dump. Now, if you know the angle of the ditch – this is something that can be determined from excavations by archaeologists – and also know the distance of the ditch from the bottom of the rampart – this again is easy enough to work out – then you can work out the best height of the rampart by using some trigonometry. A good height is one where a defender can see the attacker easily (and hence can lob a rock at them), but the attacker can only just see the defender and thus has difficulty lobbing a rock back. Indeed if d is the distance from the rampart to the ditch and θ is the angle of the ditch, then the best height h of the rampart is given by the formula

$$h = d\tan(\theta).$$

You will already have met the function $\tan(\theta)$ in Chapter 3 when we looked at sundials. If you have not read this yet then either read it, or use the tan button on your calculator.

The distance from the rampart to the ditch should also be somewhat less than the average maximum range of throwing a projectile.

Exercise. *If* $d = 10\,\mathrm{m}$ *and* $\theta = 30^\circ$, *calculate* h.

5.3 Medieval castles

We now move on 1000 years to what most of us think of as a castle, that is a Medieval castle. In England, these buildings were originally constructed by William the Conquerer following his invasion in AD 1066, and were developed greatly in later years, particularly by Edward I. The basic design of the Medieval castle remained at the forefront of military technology until the widespread use of gunpowder, and most castles in England were blown up (or at least knocked about a bit) by Oliver Cromwell following the English Civil War in the 1640s. A typical castle had to garrison a much smaller force than the hill forts, typically just one knight or lord, plus his associated troops. Many of the best examples of Medieval castles can be found in Wales. From here knights such as the legendary Sir Cumference would attempt to control the Welsh rebels under the command of their heroic leader Dai Ameter.

Motte and bailey castles

Early Medieval castles, built in England just after 1066, were called motte and bailey castles. These were miniature versions of hill forts and were built quickly by William the Conquerer to defend his new kingdom. He had to use a construction which was quick to build and very strong. To make a motte and bailey castle, a great circular ditch was dug (under duress by the local people) and using the earth from this a conical mound, called the motte, was constructed. On top of this was built a circular fortification called the keep, where the Norman lord would then live. In the earlier castles this was made out of wood and later on was constructed out of stone. Any attacker would have to climb the steep sides of the motte. At the base of the motte was a secondary wall made either of wood or stone. This enclosed an area called the bailey, where there were less vital buildings such as stables, the chapel, and the servants' quarters. A typical arrangement for a motte and bailey castle is shown in Figure 5.11.

We see again how the circle dominates the design of this castle. The reasons for this are exactly the same as in the design of the hill fort. Mathematical arguments don't change over 1000 years, even if the designers of the castles do! Examples of motte and bailey castles can be found at Restormel in Cornwall, Castle Acre in Norfolk, Totnes castle in Devon, and Oxford City Castle. The Queen of England often resides at Windsor Castle in Berkshire, which is open to the public, and the magnificent Round Tower is built on a motte with a bailey on either side.

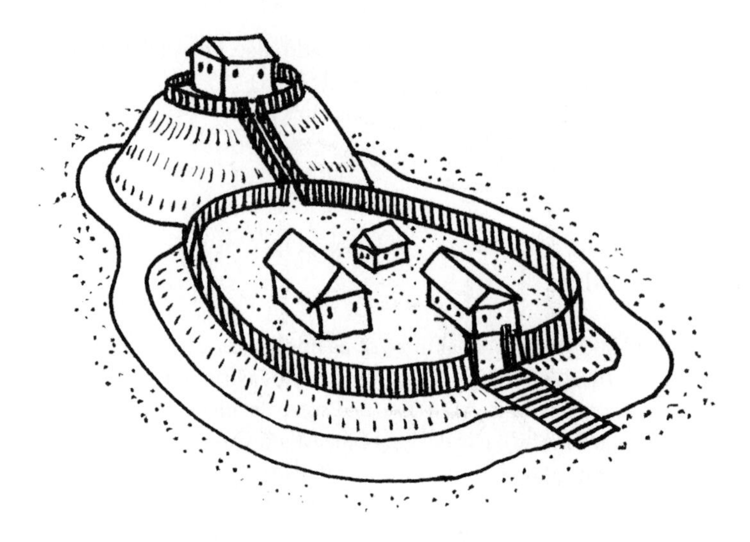

Figure 5.11: A motte and bailey castle.

Later Medieval castles

In the Eleventh century a new design of castle was developed and this is what we often think of as being a standard castle. In this design the motte was not used and instead the keep was made much larger and stronger. A typical design of a later Medieval castle is shown in Figure 5.12.

It comprised an outer defensive structure such as a ditch, a sheer cliff face, or most commonly a moat filled with water. To make the moat a more cheery place for the attackers, the castle designer usually arranged for the castle sewage to discharge straight into it. Inside the moat was an outer wall called the curtain wall. From the top of this wall the defenders would hurl rocks at the attackers or pour boiling oil on them if they got too close. Another important form of defence were archers shooting arrows both from the battlements at the top and from arrow holes in the sides. Later castles were also defended by musket men and cannon. The entrance to the curtain wall was guarded by a very strong gate-house. Inside the curtain wall was a large space called the inner bailey in which there were buildings such as the kitchen, stables, and storage areas. Finally, inside the inner bailey was the keep, which was the most heavily defended part of the castle. Much of the activity in the castle centred around the great hall which was where the main meals were eaten and the bulk of the servants slept, the lord and his lady having rather more luxurious, and secure, accommodation in the keep.

The keep was the last point of defence should the outer curtain wall be breached and typically it had extremely thick walls. Many examples of keeps survive to this day. However, the keep was just the central point in a concentric system of defences,

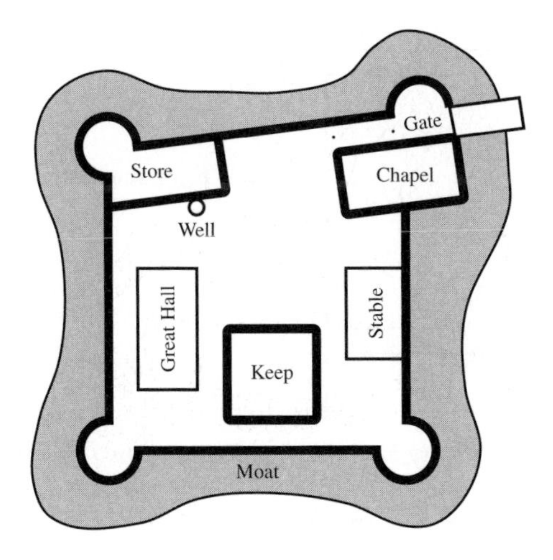

Figure 5.12: The layout of a later Medieval castle.

with several walls of defence, each designed to reinforce the other. Such castles could be effectively defended by a relatively small number of troops. They were used throughout Europe and also, in the Holy Land, by the Crusaders.

In a departure from the circular design of the motte and bailey castles the earlier keeps were square or rectangular in design. This has certain advantages for construction and living in, and straight walls are certainly easier to build than circular ones. Many castles were made with square keeps – for example the White Tower of the Tower of London. However, the departure from the 'perfect' circular form led to a far from perfect keep. In particular the corners of the keep were a major source of weakness. As we will see in Section 5.5, the corners greatly reduced the visibility of the defenders and gave a blind area where the castle could be attacked, and to overcome this, towers were built on the corners. Figure 5.13 shows a plan view of Harlech Castle in Wales which has a wonderful square keep/gate-house with towers on each corner. Harlech was built between 1283 and 1287 and it is estimated that it took about 900 people to build it and cost (at the time) $19,000.

Another weakness of the square keep is illustrated by Figure 5.14. If we draw a short line at the corner of a keep then it cuts off a large part of the corner. A line of the same length cuts off much less of a circular keep. Now suppose that this line is a tunnel. By excavating underneath the corner of a square keep it is possible to make the whole corner collapse, destroying the keep! Excavating in this way is much harder in a circular keep.

Later designs of keep became much more symmetrical to try to overcome the limitations of the square keep. Some typical examples of the plans of different types of keep are given in Figure 5.15, in which we see plans of Conigsburgh (Yorkshire),

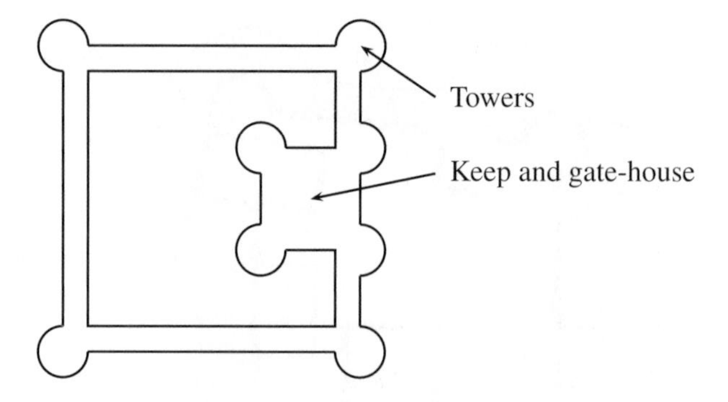

Figure 5.13: Harlech Castle plan.

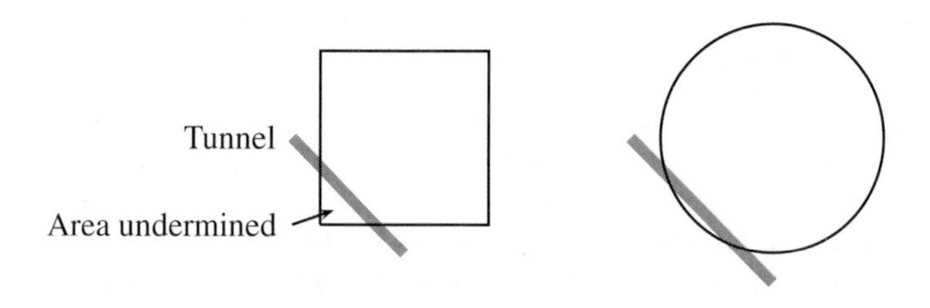

Figure 5.14: Attacking a keep.

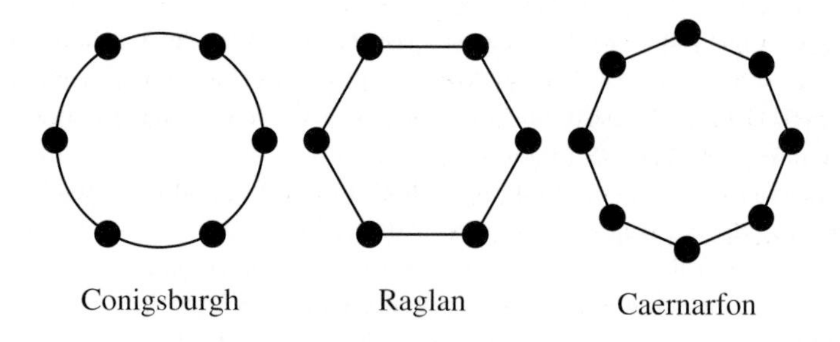

Figure 5.15: Keep plans.

Raglan (Gwent), and Caernarfon (Gwynedd) Castles. In each plan the role played by symmetry is striking. The keep at Conisburgh is circular (with six projections). The keep at Raglan is a rather stylish hexagon and that at Caernarfon an octagon. It is no coincidence that symmetry was used in this way; see the first question in the exercises.

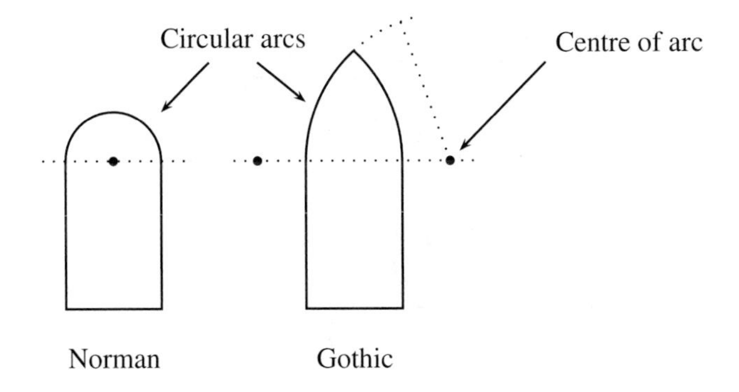

Figure 5.16: Castle window design.

Exercise. *Castles abound with many different mathematical shapes. Find as many as possible on your next trip.*

The *windows, doors, and roof beams* are in general composed of circular arcs. This gave a much larger and stronger shape than would have been possible by using straight lines. These shapes can be constructed by using a compass, and two examples of a Norman and a Gothic arch are given in Figure 5.16.

Anyone who has climbed up a castle will be familiar with the *spiral staircase*. This is a mathematical curve called a *helix*. Staircases were made in the shape of spirals as a considerable height could be achieved by a staircase that occupied very little space. Owing to the predominance amongst attackers of right-handed swordsmen, the spiral staircases always rotate clockwise when you go up. Thus, when you are climbing up the stair (and thus attacking the occupants of the castle) your right hand is always at the centre of the staircase. This makes it difficult to use your sword. On the other hand (literally) a defender coming down the stair would have their sword hand on the outside, making it much easier to use their sword.

It is possible to draw a spiral staircase (helix) as follows. For a number θ use the following formula to work out (x, y, z):

$$x = \cos(\theta), \quad y = \sin(\theta), \quad z = \theta$$

where θ is an angle and cos and sin are buttons that you can find on your calculator. The numbers (x, y, z) are the coordinates in space of a point on the helix and so taking different values of θ gives different points on the helix. Figure 5.18 shows a typical helix. Notice that the above equation leads to a anticlockwise helix.

Exercise. *How might you modify the formula to give a clockwise helix?*

An alternative way of constructing a helix, and one which closely resembles a spiral staircase, is as follows. You need a set of long thin rectangular blocks to make the stairs. The ones used for the game *Jenga* are ideal. If you don't have such a set

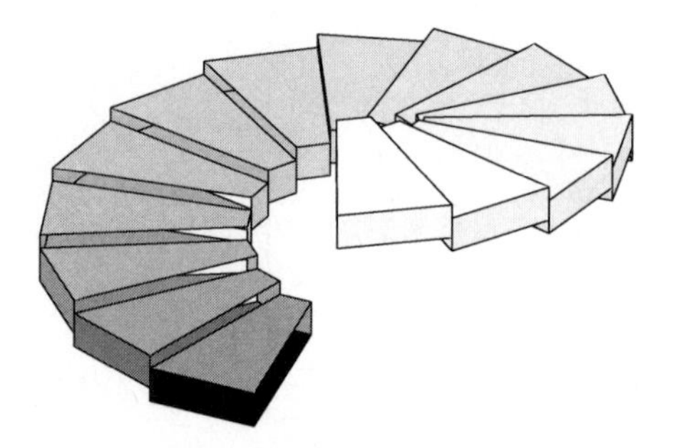

Figure 5.17: A spiral staircase.

then you can make blocks or use rectangles made out of thick card. Starting from the bottom, place one block on the next so that the diagonal of one lies over the opposite diagonal of the one below. Looking from above we would see something like this,

You will need to glue them as they won't balance. Figure 5.17 shows the finished spiral staircase. Notice that, unlike our first helix, this rotates clockwise on the way up, just like a staircase in a castle. What is happening here is that you are changing the angle of each block by a fixed amount $\Delta\theta$ each time you place one on top of the other. The blocks then automatically trace out the points (x, y) in the above formula, which goes the other way. Of course, you can easily modify the formula to make a clockwise helix.

You can also balance the blocks to form a *double helix*. Most helical staircases are single spirals but at Chambord in France, the largest and grandest of the chateau in the Loire valley, there is just such a double staircase. People believe this was designed by Leonardo da Vinci. See Figure 5.19.

5.4 More recent fortifications

With the end of the middle ages and the rise of the use of cannons and gunpowder in sieges, the design of castles changed. Indeed castles increasingly became armed forts which were part of a network used to defend a whole country. Such forts were used in

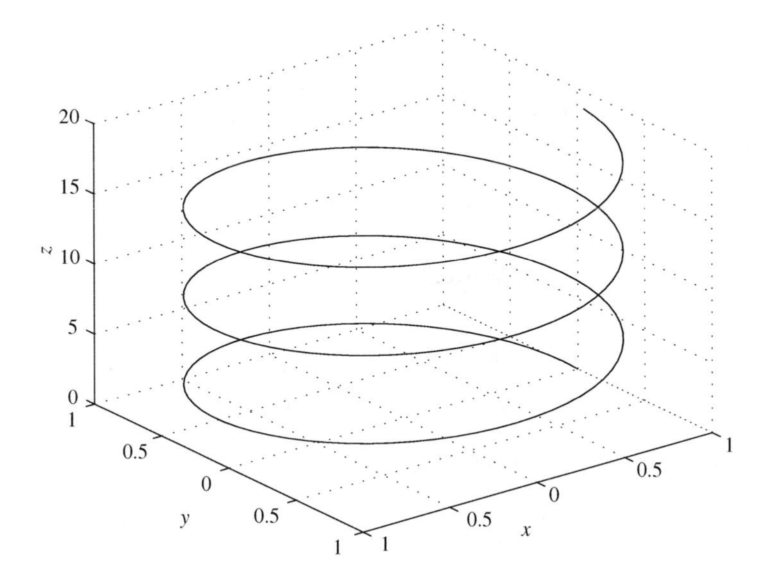

Figure 5.18: A typical helix.

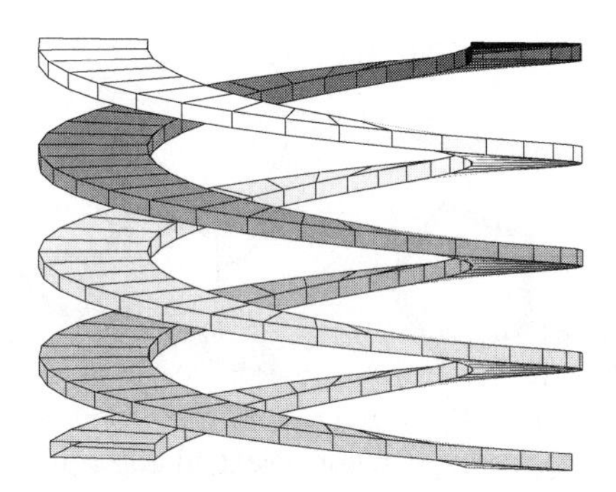

Figure 5.19: A double helix.

both world wars, but the use of air power and very long range guns has largely made them obsolete.

The design of fort which was most effective was called the *bastioned trace* and it is in these forts that we see the most striking effects of geometry. The bastioned trace was invented by the Dutch military architect Vaubain around 1530, with the aim of providing maximum defence at relatively long range using a small number of defenders armed with cannons in a fort that could be constructed easily and cheaply. The design uses symmetry and other mathematical ideas in their design to a very great extent. Nearly all bastioned trace forts are based upon a pentagonal outer wall. This was built with thick straight walls which were reinforced with rammed earth behind to absorb any shot. We will see in the next section that the vertices of the

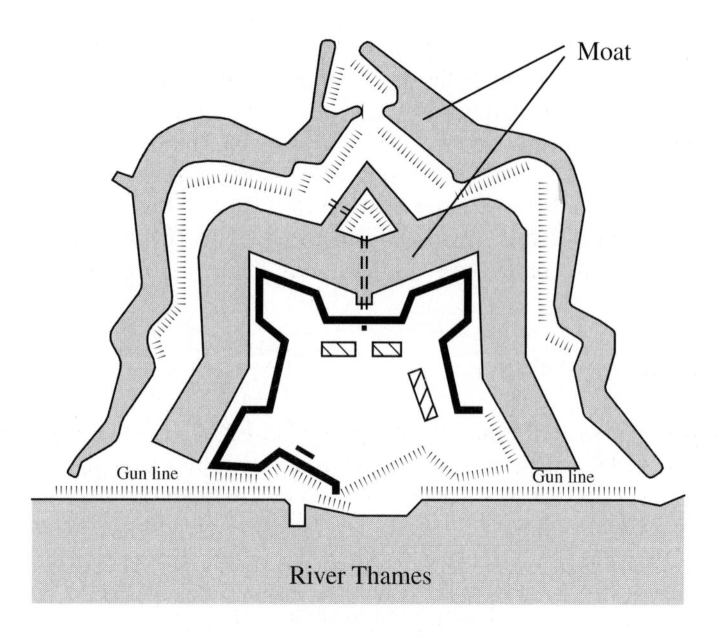

Figure 5.20: The plan of Tilbury Fort.

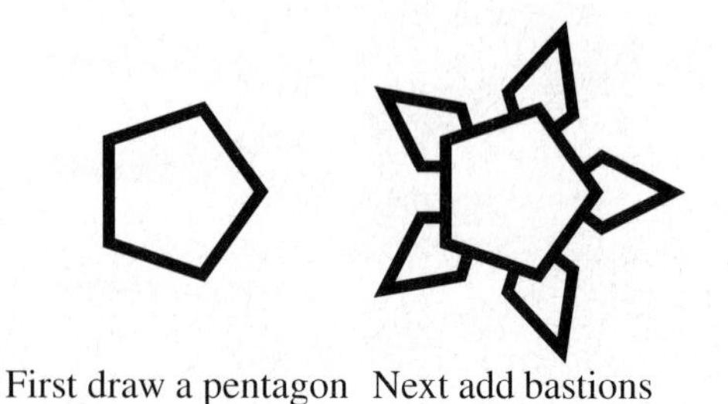

Figure 5.21: A pentagonal fort layout with bastions.

pentagon are vulnerable and to reinforce them arrow headed *bastions*, such as those shown in Figure 5.21, were constructed at each vertex. Cannon were placed in these to defend the walls of the pentagon. Inside the pentagon, and behind the earth banks, the garrison of the fort could be housed in safety and some comfort. The design of the bastioned trace was so successful that it was used for over 300 years. There are examples of bastioned trace forts all over Europe. Whole towns (such as Oxford) were defended by walls in this shape. Surviving examples of this design in the UK are Pendennis Fort in Cornwall, Portsmouth Fort, Fort George in Inverness, Charles Fort in Co. Cork, Eire, which was in use until 1920, and Tilbury Fort on the River Thames which was constructed in 1670 to defend the City of London during the wars with the

Dutch. A picture of Tilbury Fort is given in Figure 5.20. The mathematical regularity of this design is very striking. In the next section we will see the mathematical ideas leading to this essentially optimal design.

Although this design was used for the larger forts, smaller fortifications were still based upon the circle or other regular shapes. The Martello Forts used in the nineteenth century as a defence against Napolean are perfect circles, and the pill boxes used widely in the Second World War are nearly all based upon octagons. The basic design ideas which led to the construction of hill forts are still in use over 2000 years later.

5.5 How to defend a castle

Imagine you have been recruited as a designer of a castle. How should you construct it so that it is as defendable as possible with as few defenders as possible?

We will start to answer this question by thinking about an individual defender in a Medieval castle. Such a soldier would typically be armed with a longbow or a crossbow. The safest position for such a defender would be in a hollowed-out section of the curtain wall, firing through a narrow slot towards the attackers. In Figure 5.22 we see two views of such a position at Carreg Cennen Castle in Powys, South Wales.

Figure 5.22: An archer's slit at Carreg Cennen.

This position is being ably defended and we see the defender both from the inside and the outside of the castle. It is clear from these pictures that an attacker will have only a very slim chance of hitting a defender with a well-aimed projectile. From such a position the restrictions of the stonework meant that the archer would (according to our measurements at Carreg Cennen) be restricted to fire over an angle of about 45°. Furthermore the arrow from a longbow would have a typical lethal range of about 100 m depending on the weather. From such a position in a wall the defender could cover an arc of ground inside which an attacker would find life uncomfortable. However, outside the arc the attacker would be safe from the fire of this defender. This situation is illustrated below.

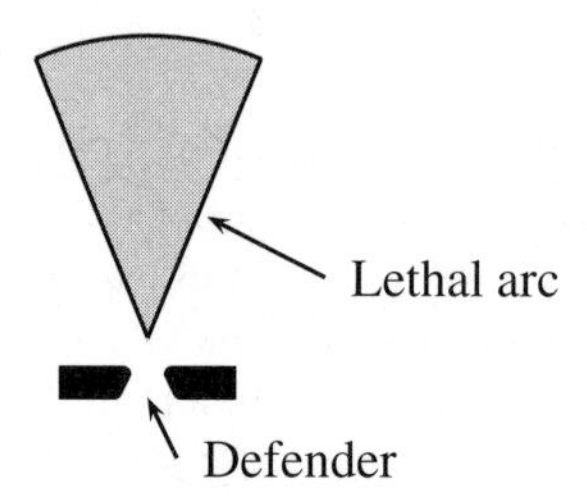

Obviously it is pointless to defend a castle with only one defender (except as a desperate last resort) and typically on a curtain wall we would see a large number of equally spaced defenders each firing through a similar defensive position. The resulting pattern of fire is shown in Figure 5.23 and it is designed so that the lethal arcs all overlap each other. Exactly the same idea was used hundreds of years later when positioning machine guns in the trenches during the First World War. In the figure you can see that any attacker must pass through an area of danger before approaching the wall. However once through this area they are relatively safe. Indeed, if they were to rush through and get close to the castle wall then they could start to attack the defences from this point by using scaling ladders, battering rams, or undermining the walls. This is clearly a weak point of the castle's defence.

Things get worse when you get to a corner of the wall in a square keep. If defenders are on the two walls meeting at the corner then there is a blind spot at the corner itself

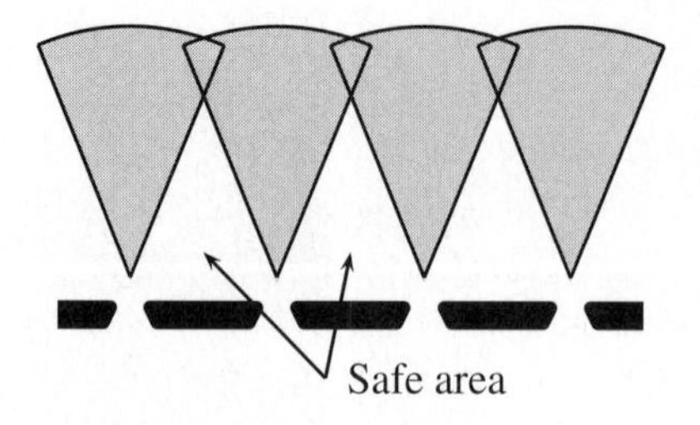

Figure 5.23: Overlapping arcs of fire.

through which an attacker can move in safety and approach the corner. Once they get to the corner they can start to excavate underneath, causing the walls to collapse.

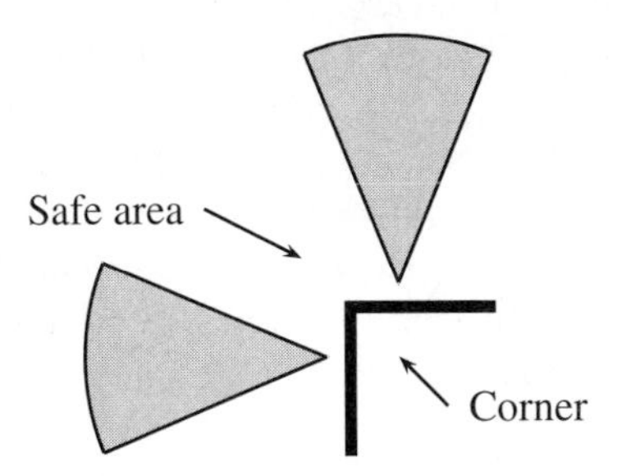

It is much easier to defend a circular keep: here there are no blind spots and all attackers need to pass through an area of danger. However there are still large areas of the walls which are difficult to defend adequately, and again an attacker is relatively safe once they get to the base of the wall.

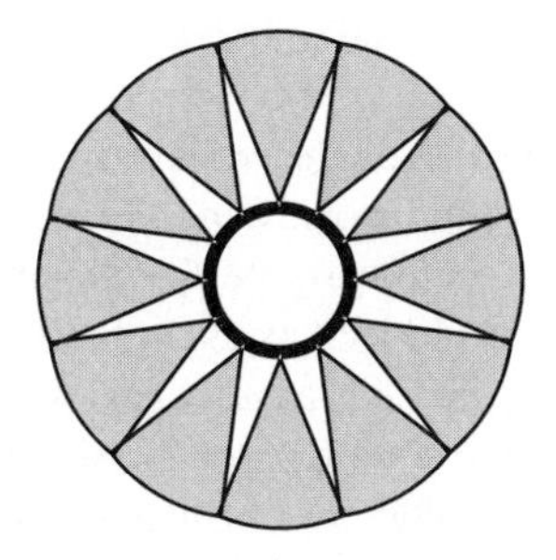

We can now pose the defenders' challenge in a more mathematical form:

> Suppose that each defender is restricted to fire over a 45° arc with a limited range (100 m) and you have a fixed (and not too large) number of defenders. What is the best shape of a castle to ensure that no part of a wall or region, in which an attacker would approach, is left undefended?

A first solution to this problem – especially effective at a corner – is to put additional circular turrets on the corners and along the walls. From these turrets defenders could then cover larger areas including parts of the original wall. The turrets had to be close enough to each other so that the areas covered by the defenders on each overlapped each other. Harlech Castle shows an effective use of turrets both on the keep and on the curtain wall. In Figure 5.24 we see how using a turret reduces the number of blind spots but does not eliminate them altogether.

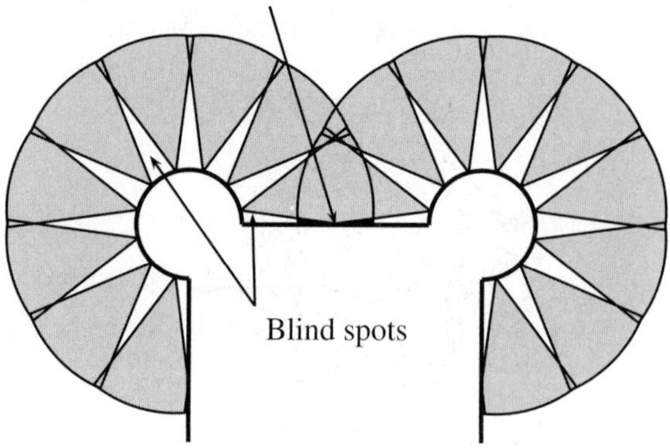

Figure 5.24: The use of turrets to reduce blind spots.

A turret is effective but still leaves blind spots. Indeed the walls of the turret itself have blind spots. A solution to this problem is to put a smaller turret on the first turret. This will then cover the blind spots on the walls of the original turret. However, even this smaller turret may have blind spots. Even better then would be to put a turret on the turret on the turret. We could carry on this process indefinitely building more turrets on more turrets. This would lead us again to the mathematical objects called fractals that we mentioned earlier. In the second set of problems you will have a chance to start drawing one when we look at the shape that you get when you put a triangle on a triangle on a triangle, etc. In practice, of course, the turrets get too small to build. But King Henry VIII of England started with a circle, built turrets on this, and then turrets on the turrets to produce a very strong design. In this design blind spots were all but eliminated and the fort had a very wide field of fire. (It is no coincidence that the design also closely resembled that of a Tudor Rose.) A wonderful example of such a fort is Deal Castle in Kent. This was built around 1540 and was part of a chain of defence which included the nearby forts at Walmer and Sandown in Kent, and which also extended to Hull and to Cornwall. These forts were all aimed to deter the Catholic alliance of Spain and France from invading the south coast and they succeeded admirably in this task. A French invading army was unable to land on the English mainland and instead ended up on the Isle of Wight where it was forced to surrender. A plan of Deal Castle is given in Figure 5.25. A recent photograph of it (together with a modern defender) is given in Figure 5.26.

Exercise. *How many circles can you see in this photograph?*

Deal Castle was a huge compound with cylindrical bastions and 145 gunports based on a six-fold symmetrical design. All of its walls were rounded to deflect shot and it was immensely strong. It was defended by powerful cannon with a range of 1 km,

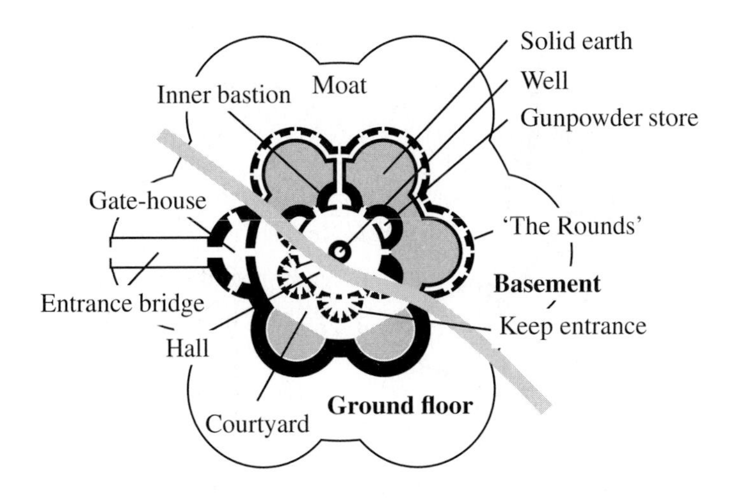

Figure 5.25: The plan of Deal Castle.

Figure 5.26: Deal Castle with a modern defender.

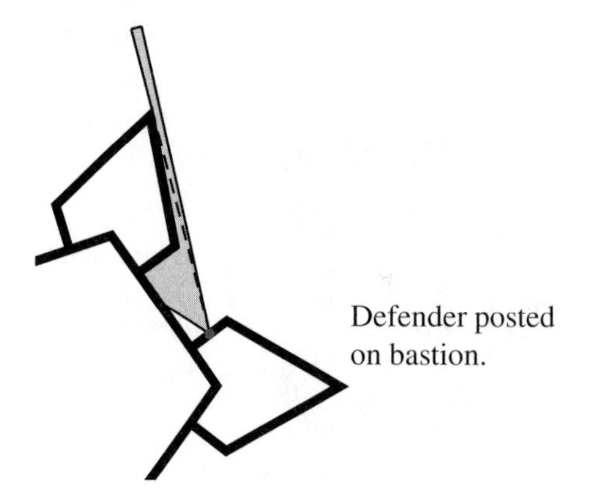

Figure 5.27: A single defender's arc of fire on a bastioned trace fort.

and there were special furnaces to heat the cannon balls so they could cause most damage to wooden ships. Deal was the jet fighter of its day and was a most formidable fortification. It is still in excellent condition despite being besieged during the Civil War and various attempts to destroy it by German bombers in the last war. It is now owned by English Heritage and is well worth a visit on a field trip.

Although very strong, Deal Castle needed around 250 troops to defend it and a lot of powder and shot for the guns. This was a large number of troops to feed and it meant that the garrison could be starved out. Exactly this happened in 1648 when the Royalist garrison surrendered to Cromwells' troops after a long seige. A further disadvantage to the fort design was that its circular walls and complex shape made it difficult and expensive to construct. As a result the design of Deal Castle was not used after the sixteenth century.

We have already seen that the more recent forts were based upon the pentagon. This is a regular polygon with five sides such that the angle between each side is 108°. It was relatively easy to build a pentagon out of a series of straight-sided walls. Starting with a basic pentagon shape, turrets in the form of pentagonal arrow-head bastions were built at each vertex. The positioning of these was so carefully thought out that each point on the wall between each bastion could be defended by a *single* gun on each of the two bastions on the vertices at the end of the wall. Furthermore, each point on a bastion could be defended by cannons firing from bastions on either side. Thus the walls of the whole fort could be completely defended by *only 10 cannon*. This is a huge improvement on the number of cannon used to defend Deal. Having defended all of your walls, the remaining cannon could then be placed on the walls of the pentagon to destroy targets at long range. You can see the effect of using bastions in Figure 5.27.

Thus we see that the bastioned trace fulfills our condition of being completely defendable with a small number of defenders. This is shown in Figure 5.28. It is *not*

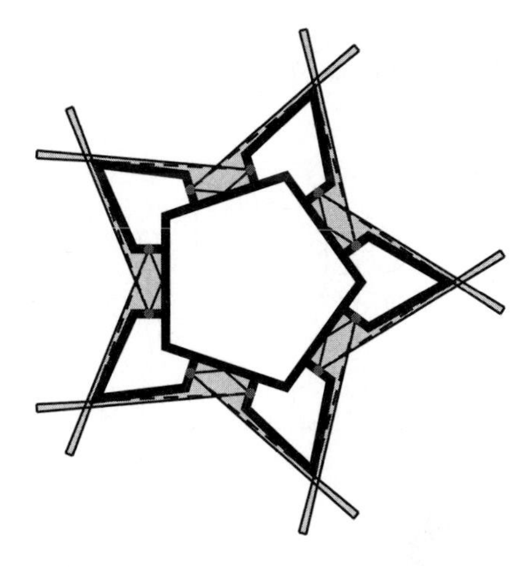

Figure 5.28: A fully defended bastioned trace fort.

the only shape with this property. In the exercises you can see whether you can come up with other shapes. However the pentagon is an excellent compromise between the needs of having a small number of defenders and a reasonably large area enclosed within the outer perimeter.

5.6 Exercises

First session

1. Many castles are symmetric. Think of as many reasons as possible why this might be a good idea.
2. A square has a perimeter of 100 m. What is the length of each side? What is its area? A circle has a circumference of 100 m. What is its radius and area? What do you notice.
3. Find the area of an equilateral triangle which has a perimeter of 100 m.
   Hint: If one side of an equilateral triangle is of length L then the area of the triangle is

$$\sqrt{3}L^2/4 = 0.43301L^2.$$

   By splitting a hexagon into six equilateral triangles, find the area of the hexagon which has a perimeter of 100 m. Place the shapes in questions 2 and 3 in order of increasing area. What do you conclude? Where would a pentagon and an octagon (eight sides) appear in this list?

4. We have seen that a circle is *convex* and *symmetric*. Which of the following shapes are convex and which are symmetric?

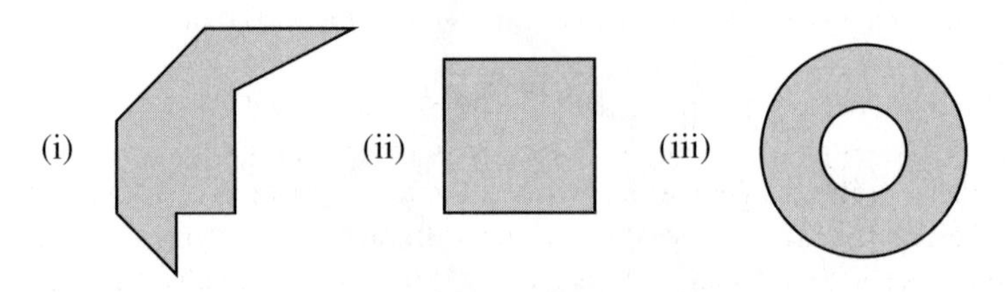

5. The entrance to Maiden Castle, illustrated in Figure 5.9, looks like this.

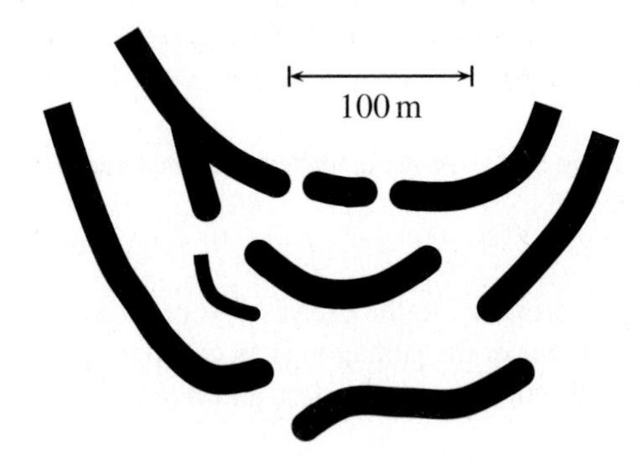

Estimate how far an attacker would have to travel to get in. Can you design a better entrance?

6. The walls of Jericho are shown in Figure 5.10. Suppose that the innermost wall has radius 1 km. Now suppose that there are four further walls built 10 m apart from one another. These will have radii 1.01 km, 1.02 km, 1.03 km, 1.04 km.
   (a) Calculate the total length of wall that an attacker would have to travel round.
   (b) Now suppose that a further five walls are built with the same separation and radii 1.05 km, 1.06 km, 1.07 km, 1.08 km, and 1.09 km. Again calculate the total length of wall that an attacker must go round.
   (c) Can you work out how far an attacker would have to go round if you build N walls with the innermost having radius 1 km and the remainder separated by 10 m? Comment on your answer.
7. Using a compass, construct a Medieval arch or a Gothic arch.
8. Using rectangles of wood, build a spiral staircase and admire its elegant helical form.
9. Ancient ditches and walls were (it is claimed) built as whole multiples of a unit of length called the 'Megalithic Yard'. At a hill fort we find a ditch of width 2.487 m and two walls of height 4.145 m and 6.632 m high. Using this data estimate the length of the Megalithic Yard.

Second session

1. Draw Deal Castle and work out the areas defended by the cannon.
2. Draw an *extended Deal Castle* by continuing the process of building turrets on the turrets.
3. For fun try building a few castles. Cardboard tubes and corrugated card make excellent towers and walls. Don't forget the spiral staircases. Experiment with constructing different geometrical shapes. Indicate on your model the effectiveness of different defensive strategies as a function of the underlying geometry. This would make a great display for open days. Who says that maths isn't fun?
4. Draw a pentagon with arrow-head bastions. Show how effectively it can be defended with only a few cannons.
5. Is a pentagon the only possible shape for a modern type of fort? Try designing bastioned forts based upon squares and hexagons making sure that each point on the fort is defended. What are the advantages and disadvantages of each of the basic shapes?

   Criteria such as number of defenders needed per area of fort defended would be a good measure here.
6. Imagine you have a long straight wall to defend with cannon that fire over a 45° arc. How might you build bastions on this to defend it? What angle to the wall should they be built at? If the cannons' range is 200 m how far should these be apart? Can you design bastions to fit round a circular wall?
7. Castle walls and tank armour are often sloped. Can you think of possible reasons for this?

Field trips

If you live in Europe, then there is no shortage of castles to visit of all ages. They all make a great trip. Further ideas for mathematical projects that involve castles can be found in the book *Maths and the Historic Environment* given in the references.

5.7 Further problems

In this section we will look at some more problems concerned with the isoperimetric theorem.

What shapes of a given area have the greatest perimeter?

This is the opposite of the isoperimetric theorem. The circle is the shape which has the least perimeter of all shapes of the same area. What shape of the same area has the greatest perimeter? We will construct a shape with a lot of symmetry but which has an *infinite perimeter* and a *finite* area.

We saw in the first problem session that amongst the triangle, the square, the hexagon, and the circle, it was the circle which had the greatest area and was the most symmetric of all of the bodies. We might be tempted to think that a highly symmetric shape will always have a large area. This is not the case. Start with an equilateral triangle with sides of length l. Now construct a *star of David* shape made by putting three equilateral triangles of side length $l/3$ onto each of the sides of the original triangle.

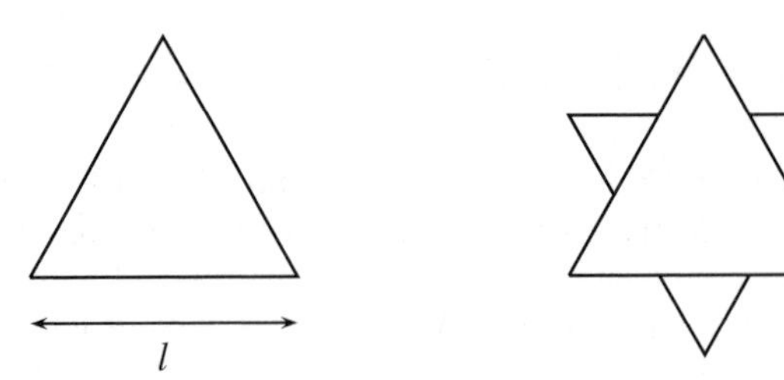

1. What symmetries does this shape have?
2. Show that the total perimeter P of the star is $4l$ and that its area A is

$$\frac{\sqrt{3}l^2}{4}\left[1+\frac{1}{3}\right]=\frac{\sqrt{3}l^2}{3}.$$

3. If $P = 100\,\text{m}$ show that $A = 360.844\,\text{m}^2$. Thus the star has a *smaller area* than the equilateral triangle of perimeter 100 m which we looked at in problem 3 of the first session.
4. We are now going to extend this idea to construct a more complicated star shape. Suppose that on the middle of each of the sides of the triangles making up the star we construct a little equilateral triangle whose side has length $l/9$. Draw this shape (which will resemble a snowflake).
5. Show that the perimeter of this new shape is $P = 16l/3$ and its area is

$$\frac{\sqrt{3}l^2}{4}\left[1+\frac{1}{3}+\frac{1}{3}\left(\frac{4}{9}\right)\right].$$

6. We can continue this idea of building small triangles on the sides of the larger ones indefinitely. Each time you produce a shape, take each side and construct on it a triangle of 1/3 the length of that side. If you continue for ever you will produce a fractal object called the *Koch snowflake*. Here is a picture of this object as it evolves:

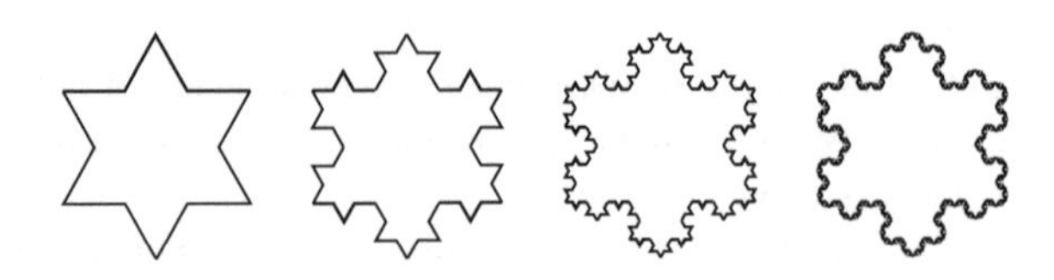

Every time we add a set of small triangles to the star its perimeter increases by a factor of 4/3. So, if we continue forever, the perimeter of the Koch snowflake must increase indefinitely and thus become *infinite*. However, the *area* A of the Koch snow flake is *finite* and we can show that it is given by

$$A = \frac{\sqrt{3}l^2}{4}\left[1 + \frac{1}{3}\left(1 + \left(\frac{4}{9}\right) + \left(\frac{4}{9}\right)^2 + \cdots\right)\right]$$

which turns out to equal $\frac{2\sqrt{3}l^2}{5}$. Thus, by constructing the Koch snowflake we have produced a shape which encloses a finite area but which has an infinite perimeter.

7. There is a family resemblance between the star of David and a Vaubain fort. However, the bastions in the Vaubain fort are small compared to the original fort and do not significantly increase its perimeter.

Proving the isoperimetric theorem

The proof has several steps. We start by making the huge assumption that there is, in fact, a shape of a given perimeter which has a bigger area than any other shape with the same perimeter. See the mathematical notes for a discussion of this point. Once we make this assumption we can then show that the shape must be a circle.

• *Step 1.* When you draw a curve you have an intuitive notion of what is inside the curve and what is outside. In the figure below we see two closed curves with the inside shaded. In the hill fort, the shaded area would be where we would live, and the unshaded part where the attackers would be.

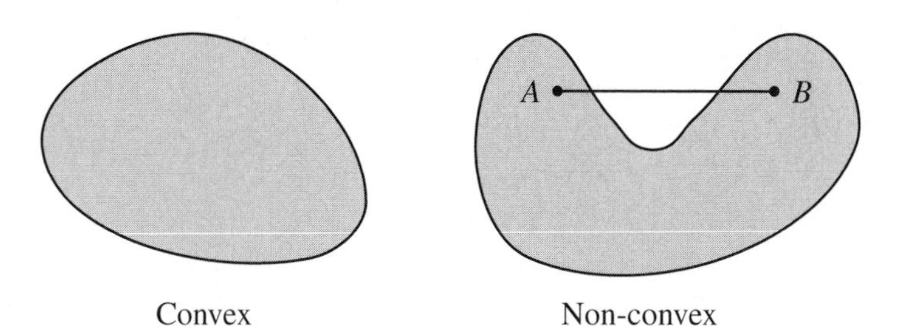

Convex Non-convex

(In fact, although we have a strong intuition about the inside and outside of a curve it is surprisingly hard to precisely define the inside and the outside of a curve, and mathematicians in the nineteenth century struggled long and hard to get a really watertight definition.)

The two shapes in the figure are fundamentally different. In the first, if you take any two points inside the curve then the straight line between them is also always inside, and we will call such a curve *convex*. In this second shape we have drawn a line from A to B which has its beginning and end inside, but passes through the curve into the outside. This shape is not convex.

We claim that a curve enclosing the most area must be convex.

It is not too hard to see why. Take a shape which is not convex and take a line from points A to B which passes into the outside at points C and D. This is illustrated below.

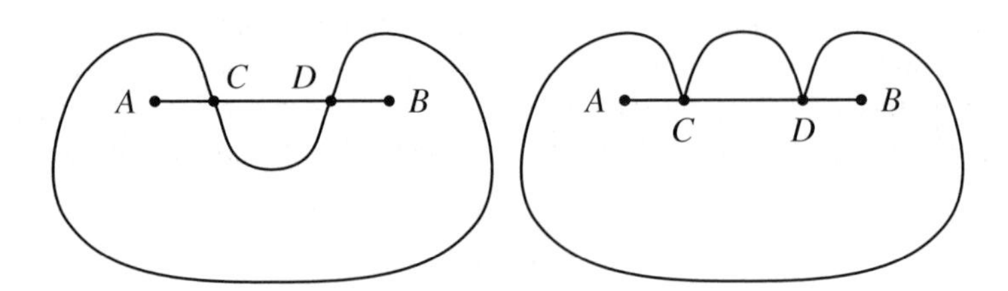

Now, take the part of the curve between C and D and *reflect it* in the line. To give the second figure above. The straight line from A to B now always stays *inside* the closed curve in the second figure, and this curve has the same perimeter as the first, as we have only reflected part of the perimeter, we have not changed its length. However, part of the region which was outside the first figure is now inside the second – so it has a bigger area. As we can do this for any line, we conclude that if a shape is not convex we can always make its area bigger by making it convex without changing its perimeter. Thus the shape with the biggest area must also be convex.

• *Step 2.* We will now show that the shape with the greatest area must be *symmetric about any diameter.*

Start with any convex shape S and take a point A on its perimeter. Now take another point B which is half-way round its perimeter. Join these two points with a straight line. We will call the line AB a diameter.

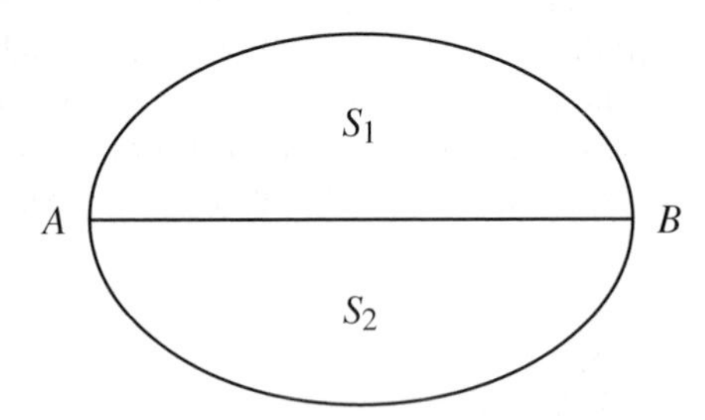

The diameter cuts the shape S into two halves, each of which has the same perimeter. Let's call these two shapes S_1 and S_2. The area of the shape S is the area of the shape S_1 plus the area of the shape S_2. Now, one of these two shapes will have a larger area than the other. Let's suppose that this is the shape S_1. If we reflect the shape S_1 about the diameter AB we get a new shape S_3 – this is drawn above. It is *symmetric* about the diameter AB. The area of the new shape S_3 is *twice* that of the area of the shape S_1, and this is bigger than the area of the shape S. However the shape S_1 *has exactly*

the same length of perimeter as the shape S. Thus, as we want to find the shape with the biggest area we would choose the symmetric shape over the shape S. Now, as the point A can be anywhere on S, we conclude that the shape with the greatest area must be symmetric about any diameter.

• *Step 3.* We now show that if A and B are two points on opposite sides of a diameter and C is any point on the curve then the angle ACB is exactly $90°$.

Let's take a shape S which we can assume is symmetric about the diameter AB and draw a point C on its perimeter. By symmetry there is another point D on the perimeter which is the reflection of C in the line AB. The shape S is then made up of a kite shape with perimeter $ACBD$ and four segments P, Q, and their reflexions P' and Q'.

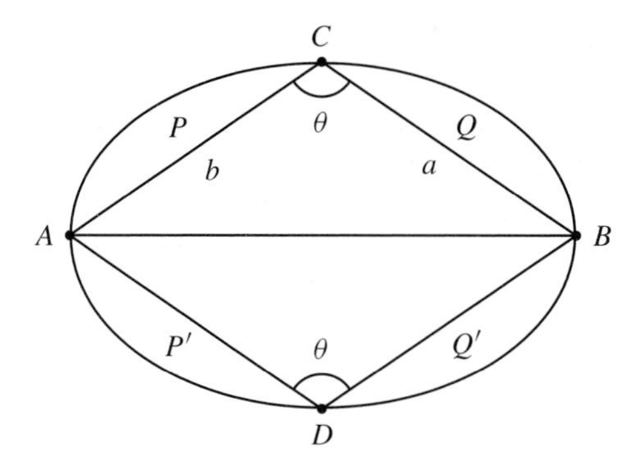

Now imagine that the segments P, Q, P', and Q' are made out of cardboard and are hinged together at the four points A, B, C and D. (You might like to try making this shape up.) By moving the segments on their hinges you can change the shape S, but, as the segments stay the same, the new shape will always have the same perimeter as the shape S. The *area* of this shape is then the area of the kite shape $ACBD$ which changes when you move the segments plus the area of the segments themselves – which does not change. We can get many different shapes by doing this and we tell them apart by measuring the angle ACB. Here are two examples:

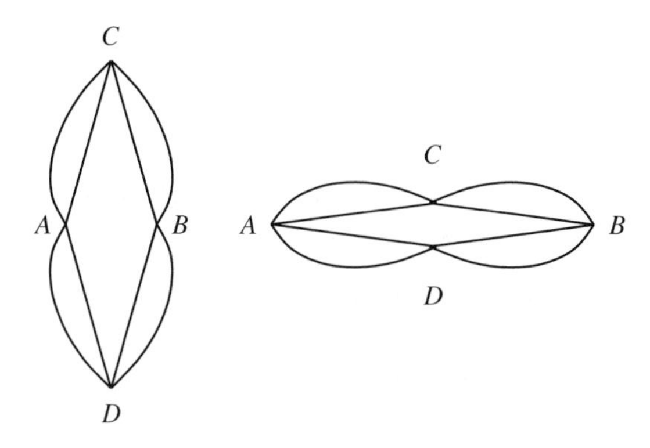

Now, here's the clever bit. If the angle ACB is called θ and the lengths of the lines AC and CB are a and b then the area R of the kite shape $ACBD$ is given by

$$R = ab\sin(\theta)$$

Here the function $\sin(\theta)$ is a standard function in trigonometry and you should be able to find the button marked sin on your calculator. All we need to know about it at present is that it takes its *maximum* value of 1 when $\theta = 90^\circ$. When we move the shape around, the lengths a and b don't change but the angle θ does. So the maximum value of R is given when $\theta = 90^\circ$. So, if we keep the perimeter length fixed and are looking for a shape with the greatest area then we should choose one in which the angle ACB is always 90°.

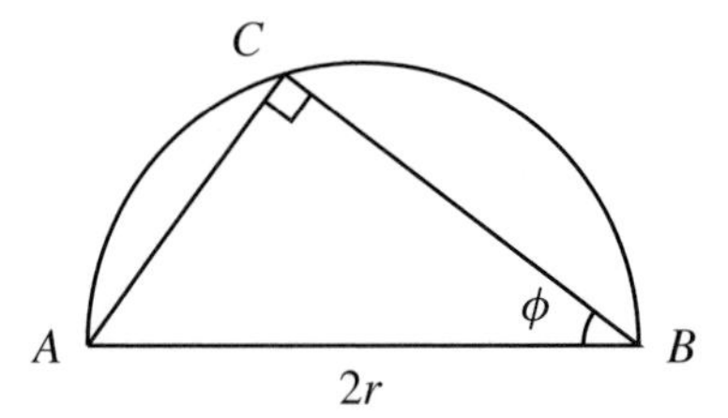

• *Step 4.* We finally show that the result we obtained in *Step 3* above means that our shape must be a circle. In the first part of the text we did this by drawing. We will now prove it. Consider a right angled triangle ACB with a right angle at C. In this triangle, the line AB is the diameter of the shape. Suppose that this has length $2r$ and the lengths of the sides AC and BC are a and b respectively. Furthermore, let the angle of the triangle at the point B be ϕ. Then from basic trigonometry

$$a = 2r\cos(\phi) \quad \text{and} \quad b = 2r\sin(\phi).$$

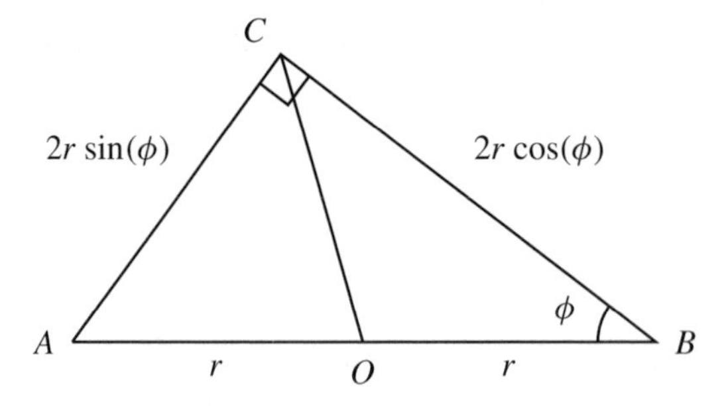

Let O be the mid-point of the line AB. The length of the line BO is r. Now look at the triangle BOC. We are going to show that the length of the line OC is also r. To do this we need to use a theorem sometimes called the *cosine theorem*.

Cosine theorem

Let BOC be any triangle with angle ϕ as shown

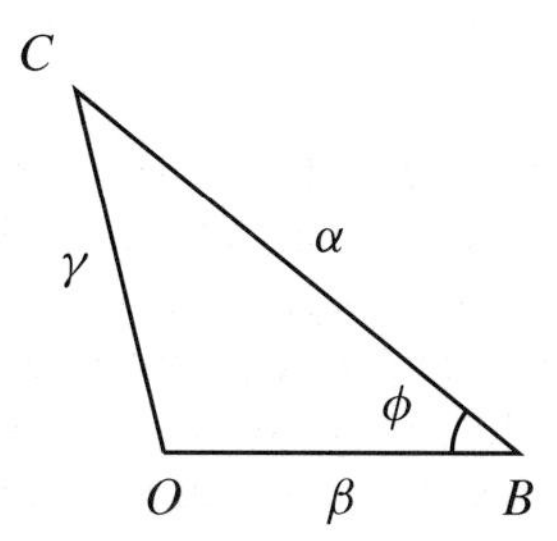

Then $\gamma^2 = \alpha^2 + \beta^2 - 2\alpha\beta\cos(\phi)$.

Suppose that BOC is any triangle with angle θ and sides of length α, β, and γ. Then

$$\gamma^2 = \alpha^2 + \beta^2 - 2\alpha\beta\cos(\phi).$$

We will now apply this to our triangle. In this case we have

$$\beta = r \quad \text{and} \quad \alpha = a = 2r\cos(\phi).$$

Thus

$$\gamma^2 = r^2 + (2r\cos(\phi))^2 - 2r(2r\cos(\phi))\cos(\phi) = r^2.$$

Hence we deduce that

$$\gamma = r.$$

What this means is that the length of the line OC is the same as the length of the lines OA and OB. This doesn't matter where C is on the outside of the shape. So the length of the point O to *any* point C on the outside of the shape is always r. This means that the shape has to be a circle, of radius r.

5.8 Answers

First session

1. Castles, keeps, and hill forts are symmetric for many reasons. These include
   - enclosing the greatest area
   - absence of weak spots
   - all-round visibility
   - ease of construction

- strength of construction
- symmetric castles look good (this is important if you are trying to impress someone)
- ease of internal communication.

Many other large constructions are symmetric, for example light-houses, gas holders, and cooling towers. Can you think why this should be so?

2. If a square has sides of length L then it has an area of $A = L^2$ and a perimeter of length $P = 4L$. Thus if $P = 100$ m we have $L = 25$ m, hence

$$A = 25^2 = 625\,\text{m}^2.$$

In contrast, if a circle has radius R then it has area $A = \pi R^2$ and perimeter $2\pi R$ where $\pi = 3.141\,592\,653\,589\,793\ldots$. Thus if $P = 100$ m then $R = 100/(2\pi) = 15.915$ m and

$$A = \pi \times 15.915\,\text{m}^2 = 795.774\,\text{m}^2.$$

We see immediately that the area of the circle is greater than the area of the square.

3. If an equilateral triangle has a perimeter $P = 100$ m then the length of one side is $L = P/3 = 33.333$ m. Using the formula in the hint gives the area as

$$A = \sqrt{3}L^2/4 = 481.125\,\text{m}^2.$$

If a hexagon has perimeter $P = 100$ m then the length of each side is $L = P/6 = 16.667$ m. Now, a hexagon is made up of six equilateral triangles, each of which has a side of length L. The area of each of these is given by $\sqrt{3}L^2/4 = 120.281\,\text{m}^2$. Multiplying by 6 gives the area A of the hexagon. This is then

$$A = 6 \times 120.281 = 721.688\,\text{m}^2.$$

If we put the shapes of the same perimeter in order of increasing area we have

$$\text{triangle} < \text{square} < \text{pentagon} < \text{hexagon} < \text{circle}.$$

If you look at these shapes you will see that this order is also in increasing symmetry. Indeed, a hexagon looks much more like a circle than a triangle because it is more symmetric. An octagon which is more symmetric than a hexagon would come between the hexagon and the circle on this list. Symmetry, however, is not the whole answer as the first of the further problems shows.

6. (a) The circumference of a wall of radius r is $2\pi r$. So, the total length of wall that the attacker will go round when all of the walls are included is

$$2\pi(1 + 1.01 + 1.02 + 1.03 + 1.04)\,\text{km} = 10.2\pi\,\text{km}.$$

   (b) In this case the distance is

$$2\pi(1 + 1.01 + 1.02 + \cdots + 1.08 + 1.09)\,\text{km} = 20.9\pi\,\text{km}.$$

Figure 5.29: The ultimate Deal Castle.

(c) To find the total distance we have to sum the arithmetic progression

$$2\pi(1 + (1 + 0.01) + (1 + 2 \times 0.01) + \cdots + (1 + (N - 1) \times 0.01))$$
$$= \pi(2 + (N - 1) \times 0.01)N.$$

Check this formula by taking $N = 5$ and $N = 10$ to obtain the answers given in parts (a) and (b). The most significant thing about this formula is that the sum grows rapidly as N increases, so we can make the attackers travel as far as we wish in range of the defences of the city.

9. Suppose that the Megalithic Yard has a length of l. Then the lengths that we find at the hill fort will be integer multiples of l. Also, the difference between any two lengths will also be a multiple of l. Taking the differences we obtain 1.658 m and 2.487 m. Now take the difference between these two. This should also be a multiple of l and we get 0.829 m. If you look you see that

$$2.487 = 3 \times 0.829, \quad 4.145 = 5 \times 0.829 \quad \text{and} \quad 6.632 = 8 \times 0.289.$$

Thus it seems reasonable from this set of data that the Megalithic Yard has length 0.829 m. Next time you visit an ancient monument, make some measurements and test out this idea.

Second session

2. Figure 5.29 shows the effect of repeatedly drawing turrets on the design of Deal Castle.
6. The following arrangement will work;

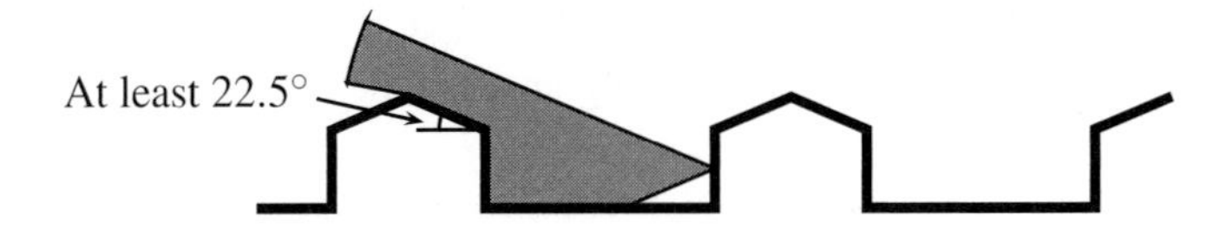

Essentially the same arrangement works when wrapped around a circular wall.

7. The base of a castle wall is often sloped into a plinth or batter. This has the shape below

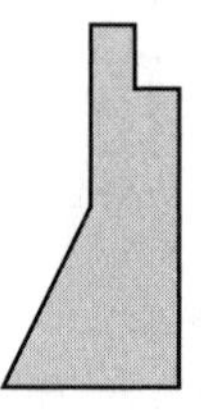

If a cannon ball is fired at this wall then the velocity of the ball at right angles to the wall is reduced. This means that the ball is less likely to penetrate the wall and may bounce off. The ball also has more of the wall to go through. The same reason applies to the side of a tank. Now, suppose that a tank is protected by armour plate of thickness t. If this sloped at an angle θ to the vertical and it is struck by a shell moving horizontally, then the thickness T of the armour which the shell has to penetrate can be worked out from the figure below

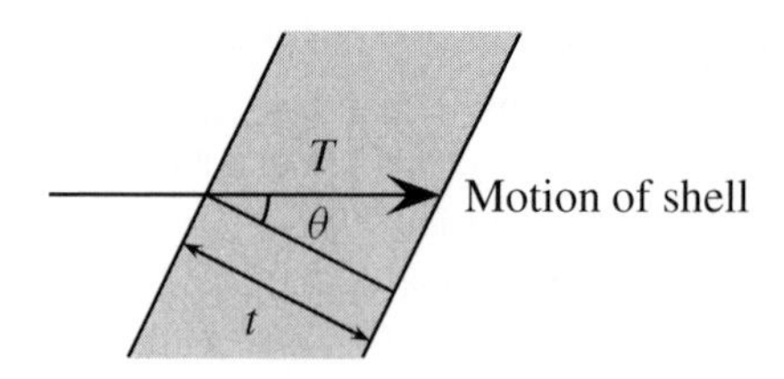

and is

$$T = t/\cos(\theta).$$

Now, whatever the value of θ we always know that $\cos(\theta) \leq 1$ so that $1/\cos(\theta) \geq 1$. In fact it is only equal to one when $\theta = 0°$ and the armour is vertical. Early tanks had vertical armour. In contrast, if $\theta > 0°$ then $T > t$ and the armour is more effective. For example, if $\theta = 45°$ then $T = 1.414t$ so we have increased the effective width of the armour by over 40%, making it much harder to penetrate.

5.9 Mathematical notes

We assumed when proving the isoperimetric theorem that there was a shape with the greatest area. Although this may seem obvious it is in fact very hard to prove. If this were false then no shape of greatest area would exist.

As a simple example of this kind of argument think of all the numbers x which satisfy

$$x > 0.$$

There are lots of these: 1, 2, $1/2$, π, etc. Is there a smallest one? If we give you a number $s > 0$ say and claim this is the smallest then you can give back $s/2$ which is also greater than zero and smaller. So we must have been wrong. Even worse whenever we claim to have a smallest number we turn out to have been wrong. Thus there is no smallest number. Can you think of an infinite sequence of numbers all greater than zero and each of which is strictly smaller than the previous one?

To give you another example consider the set R of all those rational numbers which when squared give an answer which is less than 2. The set R is perfectly well defined, however it has no greatest element. Whatever value $q \in R$ we claim to be the greatest, we can always find a larger q' which is also in R. (Of course the set R does have a least upper bound which is $\sqrt{2}$, however this is not a rational number.) It is quite conceivable that if we consider the set S of all shapes with a given perimeter then a similar situation arises. That is for any shape q we can always find another shape q' with a bigger area. Fortunately this situation does not arise, but proving that there is a shape with a greatest area is difficult. Issues like this occur very often in mathematics, particularly in the 'calculus of variations' which is all to do with finding functions which minimize certain properties, such as energy. Far from being academic, such issues are very important in understanding whether structures such as suspension bridges are stable.

5.10 References

Books and guides

More information about castles and the application of mathematics to their design can be found in the following

- Clark, R. (1985). *Castles.* Wayland.
- Copeland, T. (1992) *Maths and the Historic Environment.* English Heritage.
  This excellent book is a teachers' guide to applying mathematics in the context of historical buildings. It has many good ideas which can be developed into further projects. There are several other splendid books published by English Heritage which also aim to use the historical environment as a resource. Another good example is the following book.
- Panel, P. (1995). *Battlefields, Defence, Conflict and Warfare.* English Heritage.
- Kightly, C. (1979). *Strongholds of the Realm.* Thames and Hudson, London.
  This is the most comprehensive book that we know about the development of castles and it contains many illustrations.
- Sorrell, A. (1973). *British Castles.* Batsford.
  This book has marvellous full page pictures of castles.
- Toy, S. (1985). *Castles, their Construction and History.* Dover.
- *The I-Spy Book of Castles.* Michelin (1992).
  No British castle is missed out from the comprehensive guide which is packed full of colour photographs.

- The English Heritage guides to Deal Castle and Maiden Castle.
  There are also English Heritage guides to many other castles.
- Sanchez, S. (1979). *The Castle Story.* Penguin.
- Gleick, J. (1996). *Chaos.* Minerva.
  This book is a very gentle and interesting introduction to the modern topic of chaotic dynamics and fractals.
- Körner, T. W. (1989). *Fourier Analysis.* Cambridge University Press.
  There is a lot of literature about the isoperimetric theorem and related mathematical concepts. Much of this uses ideas from topology and is pretty advanced. The above book gives a lot of fascinating information, without swamping you in the details. It also has brilliant accounts of many other mathematical ideas, such as the development of Fourier series, the history of codes, and the applications of probability. A gourmet treat for any mathematician.

English Heritage own and manage many castles in England. Their address is:

English Heritage, Keysign House, 429 Oxford St., London, W1R 2HD, UK.
`http://www.english-heritage.org.uk`

6 How to be a spy: the mathematics of codes and ciphers

6.1 Introduction

It is not too far from the truth to say that without the use of pure and abstract mathematics there is a very good chance that the Allies would have lost the Second World War. Mathematics had many uses during the war ranging from the design of radar sets to the use of operational research to improve the way that bombers were serviced. But the most significant of all of its applications was its use in breaking the German ciphers. The main cipher used by the Germans during the war was called *Enigma* and they considered it to be unbreakable. The Enigma cipher used an automatic machine (the Enigma machine) to encipher the many complex orders issued to the German Army, Air Force, and Navy – especially the U-Boat submarines. This machine had been developed before the war and it resembled a typewriter. As a typewriter key was depressed (say for the letter E) then a light illuminated to give the enciphered version of the letter (say this is A). The design of the machine was such that if E was pressed again then a different letter (say H) would light up. It was this feature that made the Enigma cipher so hard to crack.

In the Enigma machine, as shown in Figure 6.1, there were three wheels (or in the U-boat version four wheels) with 26 contacts on each side. Each contact on one side was wired in a random manner to a contact on the other side. If a button for the letter E was pressed then an electrical contact was made sending an electric current through the first wheel. This would then be scrambled by the internal wiring and passed on to the next wheel. This would scramble it in turn for the next wheel and so on. Having passed through three (or four wheels) the path of the current was then reversed through a 'reflector' at the back of the machine, and passed through the wheels again. It would emerge fully scrambled to light the letter which was the enciphered form of E. When the next key was pressed for the next letter then the wheels moved round a bit, making the path of the current different on the next go. This is why the same letter would be enciphered in different ways on two successive presses of the same key. Every day the position of the wheels at the start of the message would be changed, making the ciphers different. Thus any attempt to break the Enigma cipher would have to start afresh each day. The Germans considered this unbreakable.

Fortunately for the Allies, a group of Polish mathematicians before the war had managed to get hold of one of the machines and determined the way that the wheels in

Figure 6.1: An Enigma machine.

the Enigma machine were wired up – a vital first start in breaking the Enigma cipher. When the war started and Poland was overrun by the German Army, the Poles passed on their secrets to the British, who assembled a team of code breakers to determine how to crack the Enigma cipher. This team had many mathematicians including Alan Turing, who was one of the greatest mathematicians in the world at that time. It also had many people who were recruited for their skill at chess or at crosswords.

The code breaking team was based in Bletchley Park near Milton Keynes. Bletchley Park is now open as a museum and well worth a trip. Because the German signals were mostly broadcast by radio they could be intercepted by radio receivers based in the south of England (without the need for spies in Germany itself), and the resulting signals were brought to Bletchley Park.

At first sight the messages were simply a random set of letters and numbers without any clues at all to their meaning. The key behind deciphering them was the fact that, although the enciphering process looked random, it obeyed strict mathematical patterns. If these patterns could be found then the Enigma cipher could be broken. To do this by hand was very slow, and it was not possible to decipher more than a fraction of the messages received.

To speed up the process of deciphering, the code breakers developed a series of machines to help them. The original machines were mechanical and were called Bombes. These machines were designed to test out many combinations of the wheels of the Enigma machine, to find a combination which would correspond to the original message. It greatly helped the deciphering process if the content of one message was known. This was then given to the Bombes which worked to find the correct wheel combinations to make this correspond to the enciphered message. To obtain such messages, aircraft would often attack certain targets, such as navigation buoys. The Germans would then report this attack, and the resulting transmission would then be intercepted. As it was known what this was about it provided a vital clue for the code breakers to work on. The beauty of this approach, was that if the combination of the wheels was worked out for one message, then the same combination could be used to crack many other messages. The information provided by this process was called Ultra and was a vital source of intelligence for the Allies throughout the war. It certainly shortened the war by many years.

The Bombes worked well for the original German ciphers. However, later on in the war a new cipher machine called the Lorenz SZ42 was introduced to encipher electrical teleprinter signals which used the international 5-bit Baudot teleprinter code. It enciphered the text by adding to it two random characters before transmission. The Lorenz machine used a complicated mechanical gearing and cam system to generate a pseudo-random sequence with very long period. This was a more complicated version of the Enigma cipher and the Germans thought this was good enough to ensure that it was unbreakable. As it was used by Hitler for his most secret messages, it was vitally important to crack it. However, it still had an underlying mathematical regularity which meant that it could be broken into. However Bombes were too slow to exploit this underlying mathematical regularity and to do this an electronic code-breaking machine was developed by Alan Turing and the electronics expert, Tommy Flowers. This machine was called Collosus and was, in a very real sense, the first electronic computer. After the war, Alan Turing went on to become one of the main pioneers of the modern computer. (Computer science students at university learn about Turing machines which are a fundamental form of computer, and about the Turing test which tries to tell a computer apart from a human being.) By using the Collosus machines the Lorenz ciphers were also broken, and the interception and deciphering of these messages gave Generals Eisenhower and Montgomery vital information prior to and after D-Day in 1944.

It should be emphasized that breaking the Enigma ciphers was always a difficult task as the arrangement of the code wheels in the Enigma machine changed every day. One of the most important uses of Enigma by the Germans was for communicating with their U-boats, which were sinking a large quantity of shipping. This threatened to reduce the United Kingdom to the point of starvation. In 1941 a Naval Enigma machine was captured when the U-boat U-110 was sunk. For a while this meant that the U-boats could be located and the convoys bringing food to the United Kingdom diverted away from them. Unfortunately, in February 1942 the German Navy introduced a new Enigma machine with four code wheels. At a stroke the code breakers at Bletchley became unable to read the U-boat messages. As a result many ships were sunk. It was only with the recovery of the code books from another sunken U-boat, U-559, that the later ciphers could be cracked. Thus, although mathematics played a vital role in the cracking of the Enigma cipher, it also required the courage of the men in the field.

6.2 What are codes and ciphers and what is the difference?

Suppose now that you have a very secret message such as `I LIKE MATHS` which you want to communicate to another friend. A secret code or a cipher is a way of disguising this message so that your friend can read it but anyone else that finds it cannot. There are many techniques for hiding a message. One is to disguise it as part of another message that does not look like a secret one. For example the message

If Lionel is kissed every morning at ten he smiles

contains interesting news for Lionel and a secret message for your friend, if they read the first letter of every word. Hiding one message inside another can sometimes be very useful. For example if you are a spy you don't want to be detected even sending a secret message.

Exercise. *Try to write a secret message concealed within another message.*

However, it is awkward to hide a very long message inside another one. Another way of disguising a message is to change every *word* for another word. This process is called a *code* and you need to have a code book to use it. The words in the original message are called *clear* words and in the coded message they are called *code* words. The code book takes you from clear to code and back again. For example you might have a code book with clear words arranged alphabetically against their code equivalent such as

Clear	*Code*
aardvark	xxcvt
anteater	aaaae
armadillo	artyu
aspidistra	qwaer
axe	cvfert

and your friend would have another code book arranged with the code words alphabetically such as

Code	*Clear*
aaaaa	cabbage
aaaab	garden
aaaac	tellytubbie
aaaad	batmobile
aaaae	anteater

These codes can be very secure but have big disadvantages. The main one is that you completely rely upon the code book which you need to transport around. If a copy of this is captured by the enemy without your knowledge, then they can easily crack any message that you send. Even if they don't get hold of the book, if you use the same book over and over again, then they quickly start to see patterns in the messages sent. For example if there are a lot of messages about *bananas* and your enemy knows this, then it won't take too long before they find out what the code word for banana is. Furthermore, codes can be very cumbersome to use. Imagine having to look up every word in a long message.

A much more popular way of disguising a message (and one particularly suited to automation) is to use a *cipher*. In a cipher you start with a *clear* message and then turn it into a secret *enciphered* message by changing it in a systematic but secret way, usually by changing or rearranging the letters in the message. There are basically two ways of doing this. The first is called a *substitution cipher* in which the *letters* of the message are changed into new ones. In the second type, called *transposition ciphers*, the letters of the message are kept the same but the order of them is changed. Both of these processes have an underlying mathematical structure and by working this out we can crack the ciphers.

6.3 The Caesar cipher

The earliest, and most famous, example of a substitution cipher was used by Julius Caesar about 2000 years ago in his campaigns (including the invasion of Britain). Caesar's idea was very simple and can best be described through an example. Suppose that we want to use a Caesar cipher to encipher a message. One way to do this is to change every letter A in the clear message to a letter E in the enciphered message. Next, every letter B in the clear message is replaced by the letter F, the letter C by G and so on, till V is replaced by Z. Then W is replaced by A, X by B, Y by C and finally, Z by D. Got that? This process can all be summarized by the following table relating the clear letters to the cipher letters.

Clear

A B C D E F G H I J K L M N O P Q R S T U V W X Y Z
E F G H I J K L M N O P Q R S T U V W X Y Z A B C D

Cipher

Using the Caesar cipher the clear message I LIKE SUMS turns into the enciphered message M PMOI WYQW.

Although Caesar only used one shift, different ciphers can be constructed by setting A to a different letter, for example J. In this case we get

Clear

A B C D E F G H I J K L M N O P Q R S T U V W X Y Z
J K L M N O P Q R S T U V W X Y Z A B C D E F G H I

Cipher

and our earlier message becomes R URTN BDVB.

There are 26 possible ways of constructing such ciphers, all of which we will refer to as Caesar ciphers, although one of these is not very mysterious (which one?) A nice way of setting up a Caesar cipher is to use a *code wheel*. This is illustrated in Figure 6.2.

Copy this picture and cut out the two wheels labled *Clear* and *Cipher*. Using a pin attach the centre of the cipher wheel to the centre of the clear wheel. To use a code

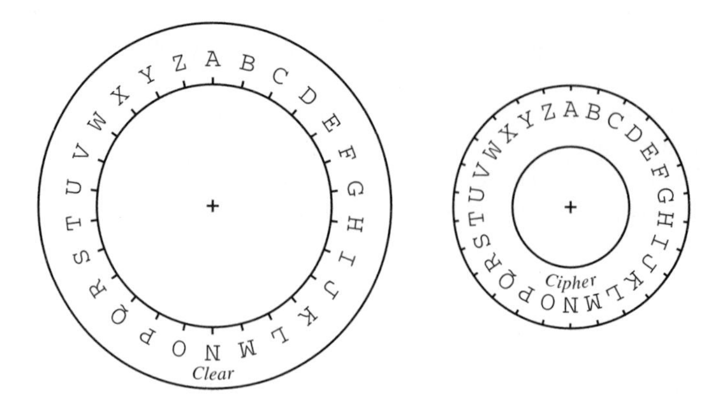

Figure 6.2: Code wheels.

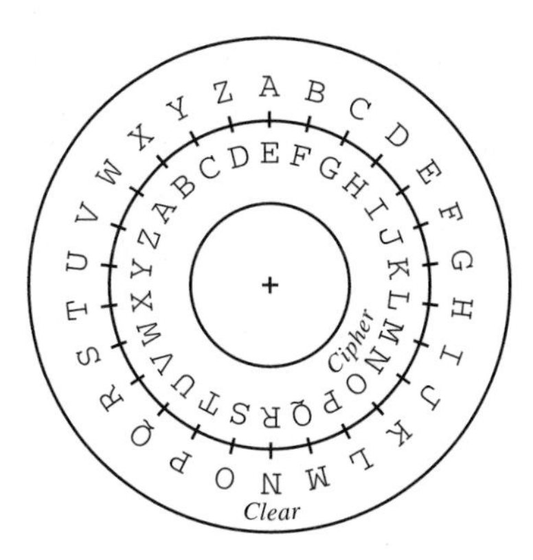

Figure 6.3: Code wheels set for the Caesar cipher.

wheel for the case of setting A in the clear message to E in the enciphered message, rotate the cipher wheel until the E on this is opposite the A on the clear wheel. Then to encipher a message take each letter in the original message and read off its enciphered version on the cipher wheel. Deciphering a message is just as easy. Every time a letter appears in the cipher read off its equivalent on the clear wheel. Try this on some of the examples in the exercises.

6.4 Was Caesar a mathematician?

We don't know. To be honest he probably wasn't, at least there is no mention of this in the classical texts. However, perhaps without knowing it, every time he used a Caesar cipher he was doing some mathematics. This mathematics is called *modular arithmetic* and is often called *clock* arithmetic. Modular arithmetic is used both in the Caesar ciphers of 2000 years ago and in the modern-day ciphers described in the section of further problems. To start with, we give each letter a number as follows

A	B	C	D	E	F	G	H	I	J	K	L	M
0	1	2	3	4	5	6	7	8	9	10	11	12

N	O	P	Q	R	S	T	U	V	W	X	Y	Z
13	14	15	16	17	18	19	20	21	22	23	24	25

Note that we have given the letter A the number 0. This may seem a bit odd at first, but it makes the subsequent mathematics much neater.

Now consider the Caesar cipher with A set to E, B set to F, etc. We will write this as

$$\mathrm{A} \rightarrow \mathrm{E},\ \mathrm{B} \rightarrow \mathrm{F},\ \mathrm{C} \rightarrow \mathrm{G},\ \mathrm{D} \rightarrow \mathrm{H}$$

and so on. Now, let's see what happens to the numbers associated with each letter. We get

$$0 \to 4, \quad 1 \to 5, \quad 2 \to 6, \quad 3 \to 7, \qquad \text{and so on.}$$

We hope that you can spot a pattern here. When the letters are converted to numbers then the process of using the Caesar cipher is nothing other than

the addition of 4.

Hold on though, things are not quite as simple as that. In this particular cipher we have `V` → `Z`, `W` → `A`, and `X` → `B`. In terms of the numbers associated with the letters this gives

$$21 \to 25, \quad 22 \to 0 \quad \text{and} \quad 23 \to 1.$$

Something interesting is happening when we get to 22. But we can make sense of this with a simple rule.

If we add 4 to the number associated with a letter then we will either get a number which is less than 26, or we will get a number which is greater than or equal to 26. In the first case we do nothing, in the second case we *subtract 26*. For example, take a clear letter `Y`; this has the numerical equivalent 24. Adding 4 gives 28, subtracting 26 gives 2. This is the numerical equivalent of `C` which is the enciphered version of Y.

Mathematicians have a name for this process. It is called

the addition of 4 modulo 26.

If x and y are the clear and cipher numbers then this is often written as

$$y = x + 4 \bmod (26).$$

There is nothing special about 4. In the case of $A \to J$ then the rule becomes $y = x + 9 \bmod (26)$. In general we can have $y = x + z \bmod (26)$, for any shift z. Using the code wheel we do this addition automatically. Modular addition is very similar to adding up numbers on a clock. For, example, if it is 10.00 a.m. now, and we go forward four hours then the time becomes 2.00 p.m. In this case moving forward four hours is equivalent to the addition of 4 modulo 12 and $2 = 10 + 4 \bmod (12)$. For this reason, modular arithmetic is often called *clock arithmetic* and we have already met this in the further problems section in Chapter 2.

Exercise. *Show that*

$$x + y \bmod (26) = y + x \bmod (26) \quad \textit{and} \quad ((x + y) + z) \bmod (26)$$
$$= (x + (y + z)) \bmod (26).$$

We can describe modular arithmetic in another way. Suppose that x is any whole number. Then we can find another number y called $x \bmod (26)$, with $0 \leq y \leq 25$. To find y knowing x, suppose first that x is greater than or equal to 26. You keep subtracting 26 from x until you arrive at a number between 0 and 25. This number is y. In other words, y is the remainder that you get when you divide x by 26. If, on the other hand, x is less than 0 then you carry on adding 26 to x until you get to a number between 0 and 25. Then this number is again y. For example

$$100 \bmod (26) = 22 \quad \text{and} \quad -15 \bmod (26) = 11.$$

Exercise. *Find* 78 mod (26) *and* −145 mod (26).

Now that we can see what the fundamental mathematics is behind the Caesar cipher, deciphering a message becomes very straightforward. As enciphering the message corresponds to addition modulo 26, then deciphering should correspond to subtraction modulo 26. Let's see how this works with our first example. Remember that A $\to$ E, B $\to$ F corresponds to $0 \to 4$ and $1 \to 5$. Thus to get from E to A we subtract 4 from 4 to get 0, and from F to B we subtract 4 from 5 to get 1. Now we also have that the enciphering process of taking Z $\to$ D and Y $\to$ C corresponds to $25 \to 3$ and $24 \to 2$. So to decipher we need to have $3 \to 25$ and $2 \to 24$. The rule in this case works as follows. Suppose that your enciphered letter has the corresponding number y and the clear letter that this corresponds to is x. If y is greater than or equal to 4, then to decipher it you subtract 4 to give x. However, if y is less than 4 then you first add 26 to y and then you subtract 4 to give x. Put in other words, we have

$$\boxed{y = x - 4 \bmod (26).}$$

Similarly, when the shift is A $\to$ J then deciphering is $y = x - 9 \bmod (26)$. This calculation can be done automatically by a code wheel.

Having defined addition and subtraction modulo 26, we can also define multiplication modulo 26 or indeed addition, subtraction, and multiplication modulo any number. Interestingly we can only really define division modulo a number if that number is *prime*, and we will look at this case in the further problems. The use of all of these operations is called modular arithmetic and it is incredibly important in code breaking from the simple Caesar cipher up to the most sophisticated ciphers in modern-day use which we will also look at in the further problems. However, you have been using modular arithmetic *all of your life* probably without knowing it. For example, suppose that today is Thursday. What day of the week will it be in 10 days time? Ask each other this – you will probably get a variety of answers until you hit on the right answer of *Sunday*. Now, suppose the days of the week are numbered so that Sunday is 0, Monday is 1, etc. In this numbering Thursday is 4. Now add 10 to 4 to give 14. There are no weeks with 14 days! To find the relevant day of the week we have to subtract off multiples of 7 (as all weeks have seven days). Doing this once we get $14 - 7 = 7$ which is too large. Doing it again we have $7 - 7 = 0$. And of course 0 corresponds to Sunday. You should spot that

$$0 = 10 + 4 \bmod (7).$$

Thus finding the correct day of the week in this case is addition modulo 7.

Exercise. *Try to think of some other examples of modular arithmetic in everyday situations.*

6.5 Statistics and more general ciphers

Suppose that you are Caesar, and you wish to encipher your secret message. It is not to your advantage to say on the message itself that A $\rightarrow$ D. If the message is intercepted then the enemy would immediately be able to decipher it. Rather better for you is to tell the recipient of the message *in advance* what the shift is going to be. Now suppose that you are an enemy agent who has intercepted a message by capturing a Roman soldier. Quite reasonably in view of where the message has come from you might reckon that it is from Caesar and uses a Caesar cipher, but you won't know the shift. How can you find this out? Well one rather tedious way is to try every possible shift. Since there are only 25 possible shifts this probably won't take very long, but it does not exploit the structure of the message, and by the time that you have deciphered the message by this process it is probably well out of date. However there is a much easier way to decipher the message which simply involves counting. If you take any message in ordinary English and count the number of times each letter appears then you will find that in almost every message the letter E appears more times than any other letter. Some other letters such as T are also common. In fact, if you take a long message and count the letters, then the most popular letters, in order of their occurrence and with the associated percentage frequency with which they occur in a 'typical' English text, are:

E	12.7%
T	9.1%
A	8.2%
O	7.5%
I	7.0%
N	6.7%
S	6.3%
H	6.1%

Languishing at the bottom are Q and Z with associated frequencies of 0.1% and 0.07% respectively.

Exercise. *Try making up a table like this yourself on any long message that you please. A good way to do this is to write down the alphabet and to put a mark against* *each letter as it comes up in the message. The message does not have to be too long to get the frequencies about right. For example, just four lines of text in this book should be enough.*

Most of the time you will come up with a similar order of letters to the table, provided that the message is long enough. (Although, be warned, this may not always happen. For example, a book about *Aaron the mathematician and an amazing aardvark* might have more than the average number of letter *a*'s.) Because all messages are different the orders of the letters for short messages will vary a bit, but these differences tend to average out for longer messages. This is an example of an important principle in the mathematical theory of probability called the *law of averages* or the *law of large numbers*. This principle is very important in statistics. If a large enough group of people is taken, then their views on what television programmes are worth watching will be close to the average views of the rest of the country.

When looking at a long message which has been enciphered, there are other clues which can help in deciphering it. If you look at single-letter words then the most common are `A` and `I` and the most common three-letter word is `THE`. This gives us a way of finding the shift in a Caesar cipher. Given a message enciphered by using a Caesar cipher, firstly count the number of times each letter occurs. For example look at the following message which has been enciphered using a Caesar cipher.

Message one

```
ZNK OJKGY UL SUJARGX GXOZNSKZOI GXK AYKJ OT LURQ JGTIKY.
ZNK SUBK UL IGYZOTM OT G ROTK UL KOMNZ OY KWAOBGRKTZ ZU
ZNK GJJOZOUT UL UTK ZU ZNK VRGIK CNKXK ZNK JGTIKX
YZGTJY, SUJARU KOMNZ.
```

In this message, the two most common letters in order are `K` and `Z` and the most common three-letter word is `ZNK`. We might guess from this that `E` $\rightarrow$ `K` and `T` $\rightarrow$ `Z`. Now, `E` $= 4$ and `K` $= 10$ so the corresponding shift is 6 letters. As `T` $= 19$ and `Z` $= 25$ the same shift works in this case as well. We conclude that this code is addition of 6 or `A` $\rightarrow$ `G`.

Exercise. *Using this shift, decipher the message. The deciphered version is given in the answers.*

The method of counting letters and setting `E` to be the most common letter doesn't always work first time. For example look at the following enciphered message

Message two

```
OZADKL EGKL WPSEHDWK GX ESRWK SJW XGMFV AF WMJGHW, AL AK
FGL GFDQ AF LZSL UGFLAFWFL LZSL LZWQ ESQ TW XGMFV. XGJ
WPSEHDW LZW SEWJAUSF AFVASFK SDKG LZGMYZL LZSL LZW
UJWLSF ESRW HSLLWJF OSK WKHWUASDDQ XAFW.
```

In this message, the letters `L` and `W` tie for first place. They can't both be the letter `E`. What may have happened in this message is that there are less `E`'s than on average and more `T`'s. Thus in principle either of `L` or `W` could be an `E` or a `T`. One way to tell which is to look for three-letter words. The most common three-letter word in plain English is THE. In the message the most common three-letter word is LZW. This is consistent with T $\rightarrow$ L and W $\rightarrow$ E but not the other way round. A little thought also shows that these two shifts also correspond to the shift A $\rightarrow$ S.

Exercise. *Now decipher the rest of the message.*

The Caesar cipher now seems very easy to crack and as a result is never used in modern ciphers. An apparently much harder cipher is a *simple substitution cipher* which substitutes another alphabet for the usual one – without the systematic shift of the Caesar cipher. Here is an example

Clear

A	B	C	D	E	F	G	H	I	J	K	L	M	N	O	P	Q	R	S	T	U	V	W	X	Y	Z
X	M	Z	R	Y	H	D	L	I	T	E	B	S	F	C	W	O	V	G	J	K	N	P	Q	U	A

Cipher

so that the message MAZES ARE AMAZING becomes SXAYG XVY XSXAIFD. It would appear, at first sight, that this cipher would be almost impossible to solve. After all there is no systematic shifting pattern which we can exploit and we can't use modular arithmetic. But in fact with a little counting it is quite straightforward to solve. Suppose that you have captured an enciphered message. You count the letters and see which is the most common. This letter is probably E and the next most common letter is probably T. Looking for the most common three letter-word (which conveniently begins with T and ends with E) then gives you the letter which corresponds to H. Looking for common one-letter words gives you A and I. This gets you going. Other letters can then be often worked out by guessing the right words. In the above example we see that T $\rightarrow$ J, H $\rightarrow$ L, and I $\rightarrow$ I. Suppose that we know this and we then find a four-letter word JLIG in the enciphered message. This deciphers to the word THI?. There are not many words like this – in fact THIS is one of them. We could guess that S $\rightarrow$ G without having to look it up. Carrying on in this manner we can break the cipher. In the appendix to this chapter are some hints about how to solve a message enciphered using a simple substitution cipher. Here is an example of such a message. Can you crack it? Hint – the words MATHEMATICS MASTERCLASS appear in it.

Message three

NOU IVIQZ SPZ ES KOQUTNBUI INOTGH UO BUUINY B FBUHIFBUECS FBSUIQCLBSS. UHI FBSUIQCLBSS PQOGQBFFI WBS SUBQUIY AZ UHI QOZBL ENSUEUTUEON KOLLOWENG UHI VIQZ POPTLBQ SIQEIS OK CHQESUFBS LICUTQIS KOQ CHELYQIN GEVIN AZ UHI FBUHIFBUECS PQOKISSOQ SEQ CHQESUOPHIQ DIIFBN. UHI KEQSU FBSUIQCLBSSIS WIQI HILY EN LONYON, ATU UHI EYIB PQOVIY SO POPTLBQ UHBU SOON UHIQI WIQI PQOGQBFFIS OK FBSUIQCLBSSIS UHQOTGHOTU UHI TNEUIY MENGYOF. UHI ABUH BNY AQESUOL GQOTP ES ONI OK UHI LBQGISU.

6.6 Multiple substitution ciphers

The very regular nature of the Caesar cipher and the fact that E is such a common letter makes such a cipher very easy to crack. The same problem occurs in the more general single substitution ciphers described above. The reason that these ciphers are so easy to crack is that in any message the letter E is always enciphered in *exactly the same way* and we can find it very easily. We can make a cipher much more secure if (like the Enigma coding machine) we encipher the letter E in different ways if it occurs twice in a message. A good way to do this is to use a *key-word*. This is a word known only to you and the person to whom you are sending the message. Key-word ciphers are really *very hard* to crack by hand and are quite secure enough for any message you might wish to send. Be warned though. Given a long enough message, a computer will make short work of a key-word cipher by exploiting its underlying mathematical regularity. We will explain how it does this at the end of this section.

An example of a key-word is the word MATHS. Suppose that the message you want to send is I THINK THAT MATHEMATICS IS GREAT. To use a cipher based upon a key-word you start by writing the key-word repeatedly above the message.

Key-word

```
M  ATHSM ATHS MATHSMATHSM AT HSMAT
I  THINK THAT MATHEMATICS IS GREAT
```

Clear

We'll start to explain how to use the key-word cipher by using modular arithmetic. To work out the enciphered message you take the numerical value of the clear letter (call this x) and the numerical value of the key-word letter directly above this letter (call this y). The numerical value of the enciphered letter (call this z) is then given by

$$z = x + y \bmod (26)$$

Got that? Well here is how it works for the first part of our message above

Numerical value y	12	0	19	7	18	12	0	19	7	18	12	0	19	7
Key-word	M	A	T	H	S	M	A	T	H	S	M	A	T	H
Clear	I	T	H	I	N	K	T	H	A	T	M	A	T	H
Numerical value x	8	19	7	8	13	10	19	7	0	19	12	0	19	7
$x + y$ mod (26)	20	19	0	15	5	22	19	0	7	11	24	0	12	14
Cipher	U	T	A	P	F	W	T	A	H	L	Y	A	M	O

so that the first part of the enciphered message is U TAPFW TAHL YAMO. In this message the first time that the letter I occurs it is enciphered as U and the second time it occurs it is enciphered as P. Because of this we can't make any progress by counting the letters in the enciphered message and looking at the most common letter. It simply will not be true that the most common letter in the enciphered message will correspond to the letter `E`. This is why these ciphers are so hard to crack by hand – provided of course that you don't know the key-word, or rather the *length* of the key-word.

How do you choose a key-word? Basically the longer the key-word the better. If the key-word is as long as the message, a set of completely random letters, and you only use it once, then provided that only the recipient of the message knows the key-word then the cipher is *totally secure*. Such ciphers were widely used by spies and were called *one-time pads*. The spy had a list of key-words (actually lists of numbers) in a pad. Every time the spy used a key-word to encipher the message the sheet from the pad was torn off and burnt. For the next message a totally different sheet was used. Using a one-time pad a spy could send short messages knowing that they could never be deciphered by the enemy. The disadvantage of a long key-word is that it becomes very inconvenient to encipher a long message. In practice you would use a shorter key-word – but not too short as the message then gets easier to decipher.

The process of using modular arithmetic to encipher a message using a key-word can get rather tedious and is greatly simplified by using what is called a Viginère table which does the sums for you. An example of such a table is given in Figure 6.4.

If you look at this square you will see that each line is an example of a shifted Caesar alphabet, with the shift increasing from top to bottom. You use the Viginère square in the same way as you would use coordinates in a graph. Suppose that the letter that you want to encipher is `E` and the key-word letter is `R`. Find the letter `R` on the left-hand side of the square above; this is marked `Key-word`. Now find the letter `E` at the top of the square where it is marked `Clear`. If you have two rulers then place one horizontally through the letter `R` and the other vertically through the letter `E`. The two rulers will cross at the letter `V`. This is the letter that you use in the enciphered message. You can now repeat this procedure with other letters in the message and other letters in the key-word. With practice you should be able to do it by eye without using the rulers at all.

Exercise. *Using modular arithmetic show why the Viginère square works.*

Using the Viginère square we can complete the process of enciphering `I THINK THAT MATHEMATICS IS REALLY GREAT` to give

```
U TAPFW TAHL YAMOWYAMPUE IL YWMLEF YDETA.
```

You can use the square in reverse to *decipher* a message. Suppose that you have received a message which is in cipher and you know that the key-word is `CODING`. Here is an example.

Message four

```
OCGMET EVDWF ZJSRZL GNZREF SCHKMZGVWFQNTU HR UNQG GHVFK
QT WPVTI GZPVIJ IS BVRN BRE UGXS VMRSGR FWZVNSWMYE
TOQLBS
```

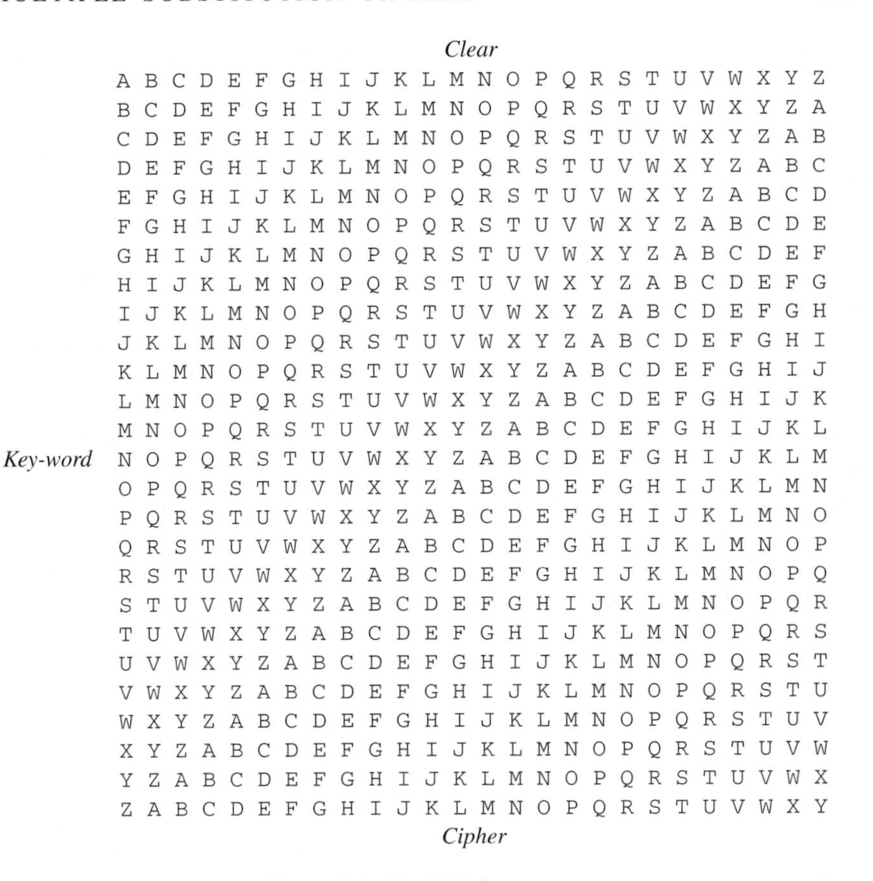

Clear

Key-word

A	B	C	D	E	F	G	H	I	J	K	L	M	N	O	P	Q	R	S	T	U	V	W	X	Y	Z
B	C	D	E	F	G	H	I	J	K	L	M	N	O	P	Q	R	S	T	U	V	W	X	Y	Z	A
C	D	E	F	G	H	I	J	K	L	M	N	O	P	Q	R	S	T	U	V	W	X	Y	Z	A	B
D	E	F	G	H	I	J	K	L	M	N	O	P	Q	R	S	T	U	V	W	X	Y	Z	A	B	C
E	F	G	H	I	J	K	L	M	N	O	P	Q	R	S	T	U	V	W	X	Y	Z	A	B	C	D
F	G	H	I	J	K	L	M	N	O	P	Q	R	S	T	U	V	W	X	Y	Z	A	B	C	D	E
G	H	I	J	K	L	M	N	O	P	Q	R	S	T	U	V	W	X	Y	Z	A	B	C	D	E	F
H	I	J	K	L	M	N	O	P	Q	R	S	T	U	V	W	X	Y	Z	A	B	C	D	E	F	G
I	J	K	L	M	N	O	P	Q	R	S	T	U	V	W	X	Y	Z	A	B	C	D	E	F	G	H
J	K	L	M	N	O	P	Q	R	S	T	U	V	W	X	Y	Z	A	B	C	D	E	F	G	H	I
K	L	M	N	O	P	Q	R	S	T	U	V	W	X	Y	Z	A	B	C	D	E	F	G	H	I	J
L	M	N	O	P	Q	R	S	T	U	V	W	X	Y	Z	A	B	C	D	E	F	G	H	I	J	K
M	N	O	P	Q	R	S	T	U	V	W	X	Y	Z	A	B	C	D	E	F	G	H	I	J	K	L
N	O	P	Q	R	S	T	U	V	W	X	Y	Z	A	B	C	D	E	F	G	H	I	J	K	L	M
O	P	Q	R	S	T	U	V	W	X	Y	Z	A	B	C	D	E	F	G	H	I	J	K	L	M	N
P	Q	R	S	T	U	V	W	X	Y	Z	A	B	C	D	E	F	G	H	I	J	K	L	M	N	O
Q	R	S	T	U	V	W	X	Y	Z	A	B	C	D	E	F	G	H	I	J	K	L	M	N	O	P
R	S	T	U	V	W	X	Y	Z	A	B	C	D	E	F	G	H	I	J	K	L	M	N	O	P	Q
S	T	U	V	W	X	Y	Z	A	B	C	D	E	F	G	H	I	J	K	L	M	N	O	P	Q	R
T	U	V	W	X	Y	Z	A	B	C	D	E	F	G	H	I	J	K	L	M	N	O	P	Q	R	S
U	V	W	X	Y	Z	A	B	C	D	E	F	G	H	I	J	K	L	M	N	O	P	Q	R	S	T
V	W	X	Y	Z	A	B	C	D	E	F	G	H	I	J	K	L	M	N	O	P	Q	R	S	T	U
W	X	Y	Z	A	B	C	D	E	F	G	H	I	J	K	L	M	N	O	P	Q	R	S	T	U	V
X	Y	Z	A	B	C	D	E	F	G	H	I	J	K	L	M	N	O	P	Q	R	S	T	U	V	W
Y	Z	A	B	C	D	E	F	G	H	I	J	K	L	M	N	O	P	Q	R	S	T	U	V	W	X
Z	A	B	C	D	E	F	G	H	I	J	K	L	M	N	O	P	Q	R	S	T	U	V	W	X	Y

Cipher

Figure 6.4: The Viginère square.

First write the key-word above the message. Now take the first letter in the key-word which is `C`. Place a ruler horizontally on `C` and follow along the ruler till you get to the enciphered letter which is `O`. Place a second ruler vertically on the letter `O` and follow the ruler up to the `clear` row of letters. You will then come to the letter `M`. This is the first letter of the original message.

Exercise. *See if you can complete the deciphering process.*

Using the Viginère square for long messages gets very tedious, as you will have discovered if you tried the above exercise. There is a much quicker way.

> Enciphering a message with a key-word can be easily done by using a computer.

The computer code exploits the fact that the process of using a key-word to encipher a message is very mechanical. In fact, we have used computer code to produce many of the examples in this chapter.

If you know the key-word then cracking one of these ciphers is easy (use the computer code). However, if you don't know the key-word then it is much harder to crack the cipher than the simple substitution ciphers we looked at earlier. In the nineteenth century ciphers of this form were thought to be unbreakable, but unfortunately for the code makers, a mathematician called Charles Babbage found that there were patterns in the enciphered message. (Babbage is also regarded as one of the early pioneers of the computer and we return to him in Chapter 8.) These patterns occurred when the same group of letters in the message coincided with the same group of letters in the key-word. In a long message this was quite likely to occur. Let's give an example. In the message `I THINK THAT MATHEMATICS IS REALLY GREAT` which we looked at earlier, the enciphered message is `U TAPFW TAHL YAMOWYAMPUE IL YWMLEF YDETA`.

If you look at the enciphered message closely you will see that the sequence `TA` appears twice, and so does the sequence `YAM`.

Now, this may be coincidence or it may be because the same bit of text is occurring underneath the same bit of the key-word. Now let's use some mathematics. If the key-word has N letters, then the distance between the occurrence of the same letters in the key word *must be a multiple of* N. Looking at the message we see that the two occurrences of `TA` are five letters apart, as are the two occurrences of `YAM`. This strongly implies that the key-word is *five letters long*. In fact the key-word is `MATHS` so we are right. This helps a great deal when trying to crack the message. Having worked out N, we now know that every Nth letter in the enciphered message is enciphered by using the same letter of the key-word. This means that every time the letter `E` is one of the Nth letters it will be enciphered *in exactly the same way*. For our example of a key-word with 5 letters, it follows that the 1st, the 6th, the 11th, and so on, letters will all be enciphered in the same way. More generally, with a key-word of N letters, the 1st, the $N + 1$st, the $2N + 1$st, the $3N + 1$st, etc., letters in the enciphered message will all be enciphered in the same way. Suppose that we were to write these letters down. If the message is long enough, then the most common letter in this sequence will be *the enciphered version of* `E`. Having worked out this we can work out the first letter of the key-word and thus decipher every Nth letter of the message. Now we repeat this process for the other letters to find each of the letters of the key-word. This process is very tedious, but with a modern computer is not too hard to implement. Indeed, another way of finding the length of the key-word is to look at the distribution of the letters separated by N other letters in the message. If N is one then this distribution will be fairly uniform, however, if N is the key-word length then the distribution will resemble that given in the table earlier, with a most common letter corresponding to `E`. This process will take a computer almost no time to implement. As a result key-word based ciphers are hardly ever used today.

6.7 Transposition ciphers

In a transposition cipher the letters of a message are not changed but their order is. A simple example of a transposition cipher is one in which you write down *every other* letter in a message till you get to the end, then you write down all the letters that you missed. For example if the message is

```
THIS CHAPTER IS ABOUT TRANSPOSITION CIPHERS.
```

then writing down every other letter gives

```
TICAT RSBUT ASOII NIHRH SHPEI AOTRN PSTOC PES.
```

Note that in the enciphered message we have arranged the letters in groups of five to make the message easier to read.

This process can be made much more general by using a code rectangle. We will illustrate this by an example. Suppose that we count the number of letters in the sentence above. This comes to 38. Now, the number 38 has only two factors which are 2 and 19 and we can write 38 as 19×2 or as 2×19. Now, suppose that we write the message as a rectangle which is 19 letters deep and 2 letters broad. This gives a pattern of the form

```
T H
I S
C H
A P
T E
R I
S A
B O
U T
T R
A N
S P
O S
I T
I O
N C
I P
H E
R S
```

If we read off the first *column*, this gives `TICAT RSBUT ASOII NIHR` (grouping the letters in fives) which you will recognize as the first half of the enciphered message. The second column then gives us the second half of the message. Now we can use the other decomposition of the number 38 to write the message in a rectangle

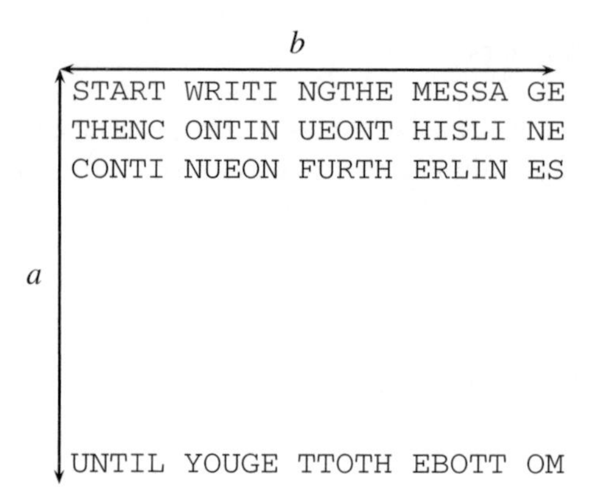

Figure 6.5: Writing text in a code square.

which is 2 letters deep and 19 letters wide. This will give a rectangle of the form

```
T H I S C H A P T E R I S A B O U T T
R A N S P O S I T I O N C I P H E R S
```

Now, if we read off each of the 19 columns then we get a new message

```
TRHAI NSSCP HOASP ITTEI ROINS CAIBP OHUET RTS.
```

Now, suppose that we have another message which has N letters. In general N will have several pairs of factor a and b such that $N = a \times b$. For example if $N = 36$ then we can express 36 as 3×12, 4×9, 6×6, 9×4, or as 12×3. Each pair of factors gives us a new rectangle and thus a new cipher. To use this write the message in an $a \times b$ rectangle which is a letters deep and b letters wide (just as in the above examples), for example if $b = 22$ then start writing the message, removing spaces, then continue until you get to the bottom. This is shown in Figure 6.5.

To encipher the message you can then write down each of the columns in turn. If you want to make the message more secure, then you can arrange with the friend who you are sending the message to, to write down the columns in a different order. Let's give an example. Suppose that our message reads as follows

```
AS WELL AS BEING ONE OF THE FOUNDERS OF THE MODERN
COMPUTER AGE, PROFFESOR ALAN TURING WAS ONE OF THE FIRST
SCIENTISTS TO APPLY MATHEMATICAL IDEAS TO THE STUDY OF
BIOLOGY. AS A RESULT OF HIS WORK WE NOW UNDERSTAND JUST
HOW THE LEOPARD GOT HIS SPOTS.
```

This message has 200 letters which is 10×20. We now write it in a 10×20 rectangle to give

1	2	3	4	5	6	7	8	9	10
A	S	W	E	L	L	A	S	B	E
I	N	G	O	N	E	O	F	T	H
E	F	O	U	N	D	E	R	S	O
F	T	H	E	M	O	D	E	R	N
C	O	M	P	U	T	E	R	A	G
E	P	R	O	F	E	S	S	O	R
A	L	A	N	T	U	R	I	N	G
W	A	S	O	N	E	O	F	T	H
E	F	I	R	S	T	S	C	I	E
N	T	I	S	T	S	T	O	A	P
P	L	Y	M	A	T	H	E	M	A
T	I	C	A	L	I	D	E	A	S
T	O	T	H	E	S	T	U	D	Y
O	F	B	I	O	L	O	G	Y	A
S	A	R	E	S	U	L	T	O	F
H	I	S	W	O	R	K	W	E	N
O	W	U	N	D	E	R	S	T	A
N	D	J	U	S	T	H	O	W	T
H	E	L	E	O	P	A	R	D	G
O	T	H	I	S	S	P	O	T	S

Now suppose that we read off the *columns* in the order 3, 1, 5, 8, 4, 2, 6, 10, 7, 9; we get the message

WGOHM RASII WCTBR SUJLH AIEFC EAWEN PTTOS HONHO
LNNMU FTNST ALEOS ODSOS SFRER SIFCO EEUGT WSHRO
EOUEP ONORS MAHIE WNUEI SNFTO PLAFT LIOFA IWDET
LEDOT FUETS TISLU RETPS EHONG RGHEP ASYAF NATGS
AOEDE EROST HDTOL KRHAP BTSRA ONTIA MADYO ETYDT

To decipher a transposition cipher we must take a different approach from the ciphers we considered earlier. The main reason for this is that, as we are using the same letters but in a different order, we will learn nothing from the statistics of the message. In fact, counting the frequencies of the letters is a good way of seeing if a transposition cipher is being used. If you count the letters in a message and find that E is the most common and T is the next most common, then the chances are that it has been enciphered by using a transposition cipher. In this case you count the number of letters in the message and look for factors of this number. Then write the message in a rectangle with sides equal to these factors. Hopefully by looking at the columns the original message will then become clear.

Exercise. *Think how you might combine a transposition cipher with a key-word base cipher to give a very secure system of enciphering a message.*

6.8 Modern-day ciphers

We will conclude very briefly with a description of how some modern-day ciphers work. These are based on deep mathematical results and we will give some more details of these in the further problems. The most successful ciphers are the so-called public-key ciphers, an example of which is the RSA cipher named after the three mathematicians Rivest, Shamir, and Adleman, who invented it. Modern ciphers are very secure if you do not disclose the key. The RSA cipher is based upon properties of prime numbers. In particular, the property that if you have two large prime numbers p and q (say 100 digits long), then it does not take very long to multiply them together to get a 200 digit number N. However, the reverse problem of finding the two prime numbers which when multiplied together give the 200 hundred digit number N is *very* hard indeed. In particular, it will take even a modern computer a very long time.

When using an RSA cipher to *encipher* a message you use the number N and a number a related to N, p and q, but not the values of p and q themselves. (For details see the further problems.) However to *decipher* the message you need to use both p and q. The person you are sending the message to chooses p and q for you, multiplies them together and sends you the number N and the related number a. In fact they can tell anyone these numbers. The incredible thing about the RSA cipher is that anyone can be told how to *encipher* a message whilst at the same time only the person you are sending messages to can *decipher* them. In fact, if you lose your copy of the plain message you might not be able to decipher the message yourself. This is quite unlike the substitution and transposition ciphers. For these ciphers, deciphering becomes very easy if you know how to encipher the message. Indeed the real problem with these methods of enciphering a message is that once the enemy guesses the enciphering process then no future message can be secure. If you send enough messages then you could assume that after a while your method of encipherment will be worked out and the code broken. In contrast, with the RSA cipher, even if the enemy works out how the encipherment process operates, the cipher remains completely secure. In an opposite manner, you can construct a cipher in which you can tell anyone you like how to *decipher* a message but only you will know how to *encipher it*. This is the way that credit cards work. The technology for reading the information on a credit card must be widely available, but only the credit card company knows how to put that information onto the card. RSA ciphers are based upon the branch of mathematics called number theory. You will meet this in most university courses on mathematics. RSA ciphers are used together with other ciphers such as the Data Encryption Standard (DES) to protect the information involved in financial transactions. Once regarded as a part of mathematics without application, number theory now lies, through its applications to cryptography, at the heart of modern technology.

6.9 Exercises

First session

1. Think of as many everyday examples of modular arithmetic as you can.
2. Devise a secret message which is hidden in another message, or which appears to be a message about something else – be creative. See if someone else can find it.
3. Make up a code wheel. Use this to decipher the following messages which were captured from a Roman soldier and are probably from Caesar himself, enciphered using the shift A → D.
   (a) SOHDVH GLYLGH JDXO LQWR WKUHH SDUWV
   (b) L DP JHWWLQJ ZRUULHG DERXW EUXWXV
4. Make up a suitably Roman message and encipher it using a Caesar cipher. See if a friend can decipher it.
5. Decipher the following messages which have been enciphered using the given shifts.
   (a) The shift A → P; decipher IWT BPNP RDJCITS JHXCV QPHT ILTCIN
   (b) The shift A → W; decipher
   W OHEZA-NQHA EO W IAYDWJEYWH YWHYQHWPKN
6. The following message has been enciphered using a Caesar cipher.

   YZUV ZKJ YVE

   Show that it could be the encipherment of two possible messages using either the shift A → R or the shift A → G.

   Can you find pairs of words which can both be enciphered using a Caesar cipher to give the same message? An example of such a pair is BOMB and HUSH.

Second session

1. The following two messages have been intercepted. We know that each of them uses a Caesar cipher but we don't know the shift in either case. Make a count of the letters to work out the shift in each case and then decipher the messages. Remember that the most common letters in English are (in order) E T A O I N S and the most common three-letter word is THE. Note – you can't guarantee that the most common letter in a message will always be E.
   (a) CQN MNBRPW XO RAXW JPN OXACRORLJCRXWB CX MNONWM CQN JWLRNWC KARCXWB OAXV CQN JCCJLT XO LJNBJA'B BXUMRNAB FJB XOCNW KJBNM DYXW J BNARNB XO LXWLNWCARL LRALUNB
   (b) DSTQJAFLZK SJW GXLWF XGMFV SK HSJL GX LZW VWKAYF GX JGESF EGKSAUK. LZW SULMSD VWKAYF AK MKMSDDQ XGMJ UGHAWK GX LZW TSKAU UJWLSF DSTQJAFLZ BGAFWV LGYWLZWJ LG XGJE S DSJYWJ ESRW
2. For what shift are the two processes of enciphering and deciphering the same?

3. Find a book or magazine in English. Count the letters in a section of it and see if the frequencies that you get correspond to those given in the table. Now repeat this exercise with a book or magazine in another language which uses the same letters as English (such as French or Spanish). See if the letters have the same order of popularity as those in English. How might this affect the way we deciphered a message from someone in France?
4. The following message was coded using the keyword RADIO. See if you can decipher it using the Viginère square.

   TOGM PIEDSWEG LA ODACQBX FXV

5. Make up a short message and encipher it using the keyword MASTER. See if a friend can decipher it.
6. The following message has been intercepted as it was broadcast – that is it comes in blocks of five letters without any breaks for the words. A spy has found that it uses the keyword SOLVING. See if you can decipher it.

   LVPVV NRWAX VBVIK IYYQN RAGMZ KBSAB RQMEE HCAPT NXABA PJYOU DWVKR YSGTO BRRDG ECMGO ESLXK HXSHP GGNTV WYQWY BWGLI IZUMB EJNNA VWPIK RVSFE DKVVS HTJV

7. Caesar's messages are being translated by the enemy, so he has decided to use a *transposition* cipher instead. The following message (which includes a dummy letter) has been intercepted from his camp. Can you decipher it?

   IICEC SORAA NEMWQ DEIUX

8. The following message contains important information, enciphered using a transposition cipher. Can you decipher it?

   IITAA TECTH CCHRA UETKS CNNNI EMLTT EODAA WIXNM SSALT PATS.

9. Make up a message and encipher it by using a transposition cipher. See if a friend can decipher it.

Field trips and projects

A fantastic field trip would be a visit to the Bletchley Park Museum. This museum includes the original mansion and many huts in the surrounding park, examples of the Enigma machines, and reconstructions of the Turing Bombes and the Collosus machine. Further museum exhibitions show the vital Polish contribution, the breaking of Japanese ciphers, and the work of the Y service in recording the signals.

The Bletchley Park Museum is managed by the Bletchley Park Trust and is close to Bletchley railway station in Milton Keynes, Buckinghamshire, UK (60 miles north of London). It is located just off the B4034, Bletchley, on the Buckingham road. Bletchley Park is open every other weekend 10.30 a.m. to 5.00 p.m. with last admissions at 3.30 p.m., and the last tour begins at 3.00 p.m.

Complete information about Bletchley park can be obtained from the Web site `http://www.cranfield.ac.uk/ccc/bpark/`

Writing computer programs to encipher and decipher a message is a fun but challenging project. Try to write some programs for yourself and test them out on the messages in this chapter.

Read about the work done by the code-breakers during the war in the books listed in the references. Other great code breaking successes were the breaking of the Japanese ciphers just before the battle of Midway and the decipherment of the Zimmerman telegram in the First World War.

6.10 Further problems

Division in modular arithmetic

Whilst it is easy to define addition, subtraction, and multiplication modulo any number, it is much harder to define division. This problem will show why, and what we have to do to define division.

(i) In normal multiplication, when we multiply two numbers together and get zero, then one of these two numbers must themselves be zero. In algebra, if $a \times b = 0$ then $a = 0$ or $b = 0$. This is not true in modular multiplication. Suppose that we are looking at multiplication modulo 26 defined over the set of numbers 0, 1, 2, 3, 4, ..., 25. Hopefully it is clear that $2 \times 13 \mod (26) = 0$. However neither 2 nor 13 are equal to 0.

(ii) Suppose that we are looking at multiplication modulo 26 and we are asked to solve the problem: find y so that $2 \times y \mod (26) = 8$. Show that $y = 4$ and $y = 17$ are both solutions to this problem. This is because $17 - 4 = 13$ and $2 \times (17 - 4) \mod (26) = 2 \times 13 \mod (26) = 0$. Remember that modular arithmetic *cannot distinguish between two answers that differ by 26*. The same problem occurs if we want to find a number y so that $2 \times y \mod (26) = 10$. Check that 5 and 18 are both solutions to this problem. Again this is happening because $2 \times (18 - 5) \mod (26) = 0$. It seems that division of any number by 2 modulo 26 is difficult because there may be two answers.

(iii) It gets worse. Can you find a whole number y satisfying $2 \times y \mod (26) = 9$. You can try but you won't find one.

(iv) Now suppose that p is a *prime number* so that it has no factors other than 1 and itself and we define modular multiplication over the set $1, 2, \ldots, p - 1$. It is not possible to have $x \times y \mod (p) = p$ because if so then one of the two numbers x and y would be factors of p.

(v) If we take $p = 13$ can we find a *whole number* y such that $2 \times y \mod (13) = 9$? At first sight this question seems rather odd. If we consider ordinary multiplication then there is no whole number solution to the problem $2 \times y = 9$. However, if you experiment you will find that $2 \times 11 \mod (13) = 9$. So modular multiplication seems very different from ordinary multiplication.

(vi) The key to understanding this result is that 13 is a prime number. Here is a remarkable result. If p is a prime number and x and z are any numbers between 1 and $p-1$ then there is a *unique* whole number y between 1 and $p-1$ such that

$$x \times y \mod (p) = z.$$

We say that multiplication modulo p of the numbers $1, 2, 3, 4, \ldots, p-1$ forms a *group*. For the definition of a group see Chapter 2. Here is an example of a *times table* for multiplication modulo 7:

×	1	2	3	4	5	6
1	1	2	3	4	5	6
2	2	4	6	1	3	5
3	3	6	2	5	1	4
4	4	1	5	2	6	3
5	5	3	1	6	4	2
6	6	5	4	3	2	1

Can you see that each number appears once, and only once, in each line and column of the table. There are many beautiful patterns in this table. Try, for example, colouring in even and odd numbers in different colours. What other patterns can you find? Compare these with the patterns that you looked at in Chapter 2.

(vii) Why does this work? Suppose that x is any number between 1 and $p-1$. Now consider the different numbers that you get if you multiply (modulo p) x by y where y lies between 1 and $p-1$. Suppose that there are two numbers y_1 and y_2 with $y_1 > y_2$ so that

$$x \times y_1 \mod (p) = x \times y_2 \mod (p).$$

This can only happen if

$$x \times (y_1 - y_2) \quad \text{is a multiple of } p.$$

Now, prime numbers have the special property that p only divides exactly into a product of two numbers if it divides one or other of the two numbers. For our problem this means that p divides exactly into either x or into $y_1 - y_2$. But as x, y_1, and y_2 are all less than p this is impossible. We conclude that if x is fixed and y takes all the values between 1 and $p-1$ then the numbers $x \times y \mod (p)$ *are all different*. (This is why the numbers in each row of the above table are all different). So as there are $p-1$ different choices for y we must get $p-1$ different numbers $x \times y \mod (p)$. So, if z is *any* number between 1 and $p-1$ there must be exactly one number y with $x \times y \mod (p) = z$. (This process is often called the pigeon-hole principle and we have already met it in Chapter 6.)

(viii) Create a 12×12 table for $p = 13$. This is an example of a Latin square and solves the Morris dancing question in Chapter 2, Second session, problem 10.

An elementary introduction to the RSA cipher

To start this description we will assume that the message that you want to encipher is not a set of words, but is instead a long *number*. This may seem a little dull to the casual reader, but in fact *all* messages are transmitted as long numbers. The usual way that a message is communicated is through a computer code which automatically transforms it into a number, through the ASCII code or some other similar mechanism. We will therefore assume that all of the information in our message can be reduced to the single number x. In fact we used this idea earlier when we transformed single letters to the numbers $0, \ldots, 25$. (Of course x will usually be a rather long number if it is to contain all of the information in the message.) The process of enciphering a message is to then apply a transformation to the number x to give a new number y. The secret of a good cipher should be that the process of finding y given x should be easy, whereas the process of finding x given y should be very hard unless you know some key to the cipher.

The building blocks of the ciphers that we will describe are *prime numbers*. Prime numbers have been investigated since the times of Euclid, but it was only recently that they became used for ciphers. Suppose that p and q are two *large* prime numbers. In practice computers take numbers with a hundred digits or so. Using these numbers we calculate $N = pq$ and take a to be *any number* which has no factors in common with $(p-1)(q-1)$. For example, if $p = 11$ and $q = 13$ then $N = 143$, $(p-1)(q-1) = 120$ and we could take $a = 7$. To encipher the message x to produce an enciphered message y we then perform the following mathematical transformation

$$\boxed{y = x^a \bmod (N).}$$

Although this calculation may look very hard, especially if x and a are large numbers, various computational tricks have been devised to make this process quick and cheap to do on a computer. Using the above numbers $a = 7$ and $N = 143$, here is a table of simple messages x and the corresponding enciphered message $y = x^7 \bmod (143)$.

x	1	2	3	4	5	6	7	8	9	10
y	1	128	42	82	47	85	6	57	48	10

You will see that there does not appear to be much regularity in the numbers y.

The number y is our enciphered message. To find y given x we only need to know N and a.

Now, suppose that we transmit the number y to a friend. How do they find x? Suppose this person knows the values of a, p, and q. Using a mathematical procedure called *Euclid's algorithm* it is easy for them to find two integers c and d so that

$$ac + (p-1)(q-1)d = 1.$$

For example, if $a = 7$, $p = 11$, and $q = 13$ then

$$-17 \times 7 + 10 \times 12 = 1$$

so that $c = -17$ and $d = 1$. Now they find a number b so that $0 < b < (p-1)(q-1)$ and $b = c \bmod ((p-1)(q-1))$. In our example we would take $b = 120 - 17 = 103$. Here is the secret of the RSA code

> To decipher the message you only need to know b and N.

In fact here is how you decipher the message

$$x = y^b \bmod (N).$$

Let's see this at work in our example by making up a table of $x = y^{103} \bmod (143)$.

y	1	2	3	4	5	6	7	8	9	10
x	1	63	16	108	125	7	123	83	113	10

If you look at the two tables you will see that if $x = 7$ then $y = 6$ and vice versa so that the procedure seems to be working in this case. Can you also see that 10 is enciphered as itself.

Why does this deciphering process work? Well the key idea behind the RSA cipher is the following result which is a version of a theorem discovered by Fermat and called *Fermat's little theorem*. (This is to distinguish it from *Fermat's last theorem*.) As with many other key mathematical ideas, Fermat discovered this theorem hundreds of years before an application was found. Suppose that p and q are any two prime numbers and x is any number which is not divisible by p or q. Then

$$x^{(p-1)(q-1)} = 1 \bmod (pq).$$

Exercise. *Try out this theorem for yourselves taking any primes p and q. For example, take $p = 2$ and $q = 3$.*

We can use this theorem to decipher our message as follows. We have chosen c and d so that $ac + (p-1)(q-1)d = 1$. Therefore we know that

$$x^{ac+(p-1)(q-1)d} = x^1 = x.$$

But, if we use Fermat's little theorem with $pq = N$ it follows that

$$x^{ac+(p-1)(q-1)d} \bmod (N) = x^{ac}x^{(p-1)(q-1)d} \bmod (N) = x^{ac} \times 1 \bmod (N)$$
$$= x^{ac} \bmod N.$$

Combining these two results

$$x^{ac} \bmod (N) = x.$$

But, $b = c + k(p-1)(q-1)$ for some integer k so that, again using Fermat's little theorem

$$x^{ac} \bmod (N) = x^{ab}x^{-k(p-1)(q-1)} \bmod (N) = x^{ab} \bmod (N) \times 1 = x^{ab} \bmod (N).$$

This shows us that

$$x^{ab} \bmod (N) = x.$$

Now, when we encipher our message we produce the message y given by $y = x^a \bmod (N)$. Therefore, from the expression above we can see that

$$y^b \bmod (N) = x^{ab} \bmod (N) = x.$$

Aha, this means that given b and N we can find x from y.

Exercise. *Now try the enciphering and deciphering processes out for yourselves using suitably chosen prime numbers p and q and a message x. Hint: Don't make p, q, or x too large.*

6.11 Answers

Exercises in the text

Message one

```
THE IDEAS OF MODULAR ARITHMETIC ARE USED IN FOLK DANCES.
THE MOVE OF CASTING IN A LINE OF EIGHT IS EQUIVALENT TO
THE ADDITION OF ONE TO THE PLACE WHERE THE DANCER STANDS,
MODULO EIGHT.
```

Message two

```
WHILST MOST EXAMPLES OF MAZES ARE FOUND IN EUROPE, IT IS
NOT ONLY IN THAT CONTINENT THAT THEY MAY BE FOUND. FOR
EXAMPLE THE AMERICAN INDIANS ALSO THOUGHT THAT THE CRETAN
MAZE PATTERN WAS ESPECIALLY FINE.
```

Message three

The letter frequencies in the encoded message are:

A	5	B	29	C	10	D	1	E	20	F	15
G	9	H	22	I	48	J	0	K	8	L	15
M	1	N	18	O	31	P	11	Q	34	R	0
S	37	T	11	U	42	V	4	W	4	X	0
Y	10	Z	6								

The most frequent letter is I. The most frequent bigrams are:

bigram	UH	HI	SU	IQ	FB	BS
frequency	15	14	13	12	10	10

The word UHI appears 9 times and the bigram *SS* six times. We used the substitution cipher

Plain letters: ABCDEFGHIJKLMNOPQRSTUVWXYZ
Coded letters: BACYIKGHEJMLFNOPRQSUTVWXZD

The plain text message is

NOT EVERY SPY IS FORTUNATE ENOUGH TO ATTEND A MATHEMATICS MASTERCLASS. THE MASTERCLASS PROGRAMME WAS STARTED BY THE ROYAL INSTITUTION FOLLOWING THE VERY POPULAR SERIES OF CHRISTMAS LECTURES FOR CHILDREN GIVEN BY THE MATHEMATICS PROFESSOR SIR CHRISTOPHER ZEEMAN. THE FIRST MASTERCLASSES WERE HELD IN LONDON, BUT THE IDEA PROVED SO POPULAR THAT SOON THERE WERE PROGRAMMES OF MASTERCLASSES THROUGHOUT THE UNITED KINGDOM. THE BATH AND BRISTOL GROUP IS ONE OF THE LARGEST.

Message four

MODERN CHAOS THEORY ALLOWS MATHEMATICIANS TO MAKE SENSE OF THINGS WHICH UP TILL NOW HAVE SEEMED COMPLETELY RANDOM

First session

1. Examples of modular arithmetic include: the 12 hour clock, the 24 hour clock, the days of the week, the days of the year, the days of the month (although note that the length of each month varies so you don't always use the same number when doing calculations), anything involving remainders such the number of coins that you get when you convert a sum of money into coins and notes.
3. (a) PLEASE DIVIDE GAUL INTO THREE PARTS
   (b) I AM GETTING WORRIED ABOUT BRUTUS
5. (a) THE MAYA COUNTED USING BASE TWENTY
   (b) A SLIDE-RULE IS A MECHANICAL CALCULATOR

6. The message deciphers to either of the two messages `HIDE ITS HEN` or `STOP TED SPY`. Other examples of pairs are `MUNCH` and `SATIN` or `PECAN` and `TIGER`.

Second session

1. (a) This message has the shift A $\rightarrow$ J

   THE DESIGN OF IRON AGE FORTIFICATIONS TO DEFEND THE ANCIENT BRITONS FROM THE ATTACK OF CAESAR'S SOLDIERS WAS OFTEN BASED UPON A SERIES OF CONCENTRIC CIRCLES

   (b) This message has the shift A $\rightarrow$ S

   LABYRINTHS ARE OFTEN FOUND AS PART OF THE DESIGN OF ROMAN MOSAICS. THE ACTUAL DESIGN IS USUALLY FOUR COPIES OF THE BASIC CRETAN LABYRINTH JOINED TOGETHER TO FORM A LARGER MAZE

4. CODE BREAKING IS AMAZING FUN
6. THE ANALEMMATIC SUNDIAL IS BECOMING VERY POPULAR IN PUBLIC PLACES AS IT TELLS THE TIME ACCURATELY AND INVOLVES AN AMOUNT OF AUDIENCE PARTICIPATION
7. I CAME I SAW I CONQUERED
8. I CAN'T WAIT UNTIL THE NEXT ACTION PACKED MATHS MASTERCLASS

6.12 Mathematical notes: Some properties of the English language

It is very helpful when trying to crack a simple substitution cipher to know some statistical properties of the English language. We have already looked at the relative frequency of the letters, and know that THE is the most common three-letter word. Here are some other properties. The following facts about English might be useful when trying to crack a code.

All lists are given in order of most common to least common.

Most common letters

1. E
2. T
3. A, O, N, R, I, S
4. H
5. D, L, F, C, M, U
6. G, Y, P, W, B
7. V, K, X, J, Q, Z

Most common pairs:
TH, HE, AN, RE, ER, IN, ON, AT, ND, ST, ES, EN

Most common double letters:
LL, EE, SS, OO, TT, FF, RR, NN, PP, CC, MM, GG

Most common triples:
THE, ING, CON, ENT, ERE, ERS, EVE, FOR, HER, TED

Most common:

initial letters:	T, A, O, M, H, W, C, I, P
second letters:	H, O, E, I, A, U, N, R, T
third letters:	E, S, A, R, N, I
final letters:	E, T, S, D, N, R, Y, G

It is also significant that more than half of English words end with the letter E.

6.13 References

There are many books on codes and ciphers, lots of which describe the underlying mathematics. Here are just a few

- Gardner, M. (1984). *Codes, Ciphers and Secret Writing.* Dover.
- Lewin, R. (1978). *ULTRA Goes to War, the Secret Story.* Book Club Associates.
- Winterbotham, F. W. (1974). *The Ultra Secret.* Wiedenfeld and Nicolson.
- Smith, M. (1998). *Station X: The Code Breakers of Bletchley Park.* Channel 4.
- Singh, S. (1999). *The Code Book.* Fourth Estate.
  This book gives the history of code breaking from ancient until modern times. The same author also wrote the fantastic book *Fermat's Last Theorem* which is well worth reading by anyone interested in mathematics and its history.
- Hodges, A. (1992). *Alan Turing: The Enigma.* Vintage.
  As well as describing Turing's work on code breaking, this book also gives an insight into one of the greatest mathematical minds of the twentieth century. This book has also been produced as a play and as a TV documentary.
- Baker, H., and Piper, F. (1982). *Cipher Systems, The Protection of Communications.* Wiley.

The Web site http://members.aol.com/nbrass/biblio.htm gives a very large list of interesting books and references related to code breaking.

7 What's in a name?

7.1 Introduction

Before we get stuck into this chapter ask around and try to find out what a number is. The kind of response will probably range from 'Well it's obvious, everyone knows what a number is' all the way to 'Hmm, well now you come to mention it, I'd have to think carefully a bit before I give you an answer'. If possible ask your nearest friendly mathematician. When stuck with a question like this, a good way to answer it is to look in a dictionary. A dictionary will give you an answer something like this,

> 'Number':
>
> - a quantity, amount, or the total count;
> - a symbol representing this.

The dictionary seems to be telling us that a number is two things which are quite different because the way we choose to write numbers and the number itself are *not* the same. Historically, as we will soon see, there have been many different ways of representing the same quantity. As mathematicians we use numbers every day and thinking for a while about how we write them is not only interesting but also quite fun.

Not all cultures have been able to count even though we take it for granted. For example some Australian aboriginal tribes only had words which described the quantities 'one', 'two', and 'many'. The Botocoudo Indians in Brazil had words for amounts up to four but then had a word which means 'many'. When you say this word you always point to your head to indicate that there are lots of things, like the hairs on your head and you can't count that far. We all know how to count things: one, two, three, four, . . . , etc., so we can get started and look at how different people have chosen to talk about numbers and in particular how to write them down.

We have already used a system to write down numbers. It's plain language: 'one', 'two', etc. This isn't good if you are a mathematician because it is long winded. Every time you want to do a sum you have to do it in your head. For example, how would you divide six hundred and forty nine by eleven? (The answer is, of course, fifty nine).

One good way to help remember things is to find patterns. Finding such patterns is one major part of what makes up mathematics. Do you know a pattern that helps you remember your nine times table? Lets write out the first part of this:

$$\begin{aligned} 1 \times 9 &= 9 \\ 2 \times 9 &= 18 \\ 3 \times 9 &= 27 \\ 4 \times 9 &= 36 \\ 5 \times 9 &= 45 \end{aligned}$$

Look at the numbers on the right. Try adding the two digits together. So $1 + 8 = 9$, $2 + 7 = 9$, and $3 + 6 = 9$. There seems to be a clear pattern here. In fact we have already met this pattern in Chapter 6. When you add the two digits on the right you get 9 – the times table we are calculating! Such patterns are much harder, if not impossible, to find if numbers are written out with words.

Exercise. *Using this fact can you work out* 6×9*?*

Exercise. *Does this pattern hold for all multiples of* 9*? If not can you find a counterexample? (Hint: have a look at Chapter 6.)*

Some ancient number systems

The simplest and perhaps oldest way of writing numbers is a tally system. The oldest object which people believe has mathematical meaning is a wolf bone found in Czechoslovakia which dates from about 30,000 BC and has 57 deep notches cut into it. One side of the bone is shown in Figure 7.1. Whether this bone has meaning, and if so what, isn't known. Many of the lines appear to be grouped in fives. Could this be the tally of a hunter's kills perhaps?

Tally sticks were used in Britain as tax and accounting records as late as 1828. When this system was abandoned there was a huge pile of sticks left over. In 1834 it was decided to get rid of them in a bonfire. Unfortunately this got a bit out of hand and the parliament buildings burnt down as well! The drawback of tally sticks is that

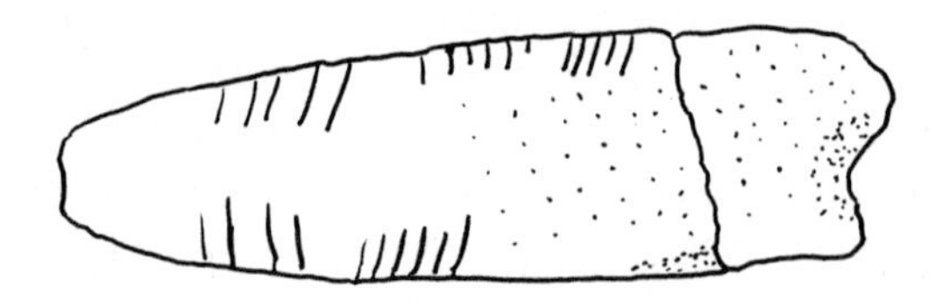

Figure 7.1: A sketch of the Czechoslovakia wolf bone.

arithmetic gets cumbersome very quickly as soon as you have large numbers. When this happens you can do one of two things:

- introduce different kinds of counters to represent different amounts;
- place counters in different positions. How much a counter represents depends on where you put it.

In fact we still use the first idea for money. We have bank notes worth different amounts.

Ancient Egypt

Egyptian hieroglyphic writing used different symbols for different numbers. The order in which you write them down doesn't matter although traditionally you write down the larger ones first. We know about Egyptian mathematics because of a special document known as the Rhind papyrus. This was discovered by the Scottish Egyptologist Alexander Henry Rhind who purchased it in Egypt in 1858. It was written by a scribe called Ahmes around 1650 BC and contains a number of worked problems including fractions and gives methods for solving various equations. The papyrus is now kept in the British Museum.

Some of the number symbols used are shown in Figure 7.2. To write a number using this system you simply use as many symbols as you need. So 'thirty seven' would be three 'ten' symbols and seven 'ones'. The order in which you write them doesn't matter but it was important to the ancient Egyptians that they were arranged neatly on the page. Interestingly, Egyptian hieroglyphs were written in *both* directions: both from left to right (as we do in English) and from right to left. This was to make their work more symmetrical. You can easily work out which way to read a particular piece of text by looking at the way living things face. They always face you as you approach them. This type of system is limited because you keep needing to invent different symbols.

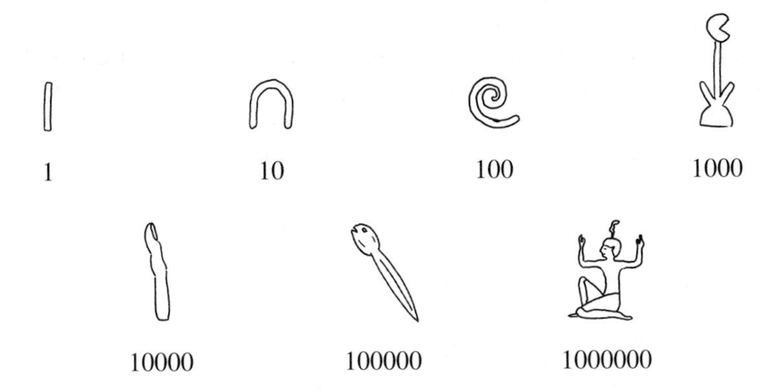

Figure 7.2: Some Egyptian number symbols.

I one
V five
X ten
L fifty
C one hundred
D five hundred
M one thousand

Figure 7.3: Roman numerals.

Exercise. *What does the number* 𓆼 𓆼 𓎆 𓏺 𓏺 𓏺 *mean?*

Exercise. *Write down* 124037 *in Egyptian hieroglyphs.*

Roman numerals

The Romans also used different symbols, some of which are shown in Figure 7.3.

Originally the ancient Romans used as many of these symbols as they needed to write a particular number. For example 'fourteen' would be written as XIIII.

This differs from Roman numerals as we know them now. Later they developed a system which used both the symbol *and* its position to give its value. For example they began to write four as IV instead of IIII as they had before. In particular IV became different from VI.

At its height the Roman Empire ruled the whole of Europe and many areas in what is now Egypt, North Africa, and the Middle East. Despite their successes the Romans never developed much mathematics of their own. One theory suggests that this is because they were held back by their comparatively poor grasp and use of numbers.

Exercise. *Write down the following numbers in Roman Numerals: (i) 7, (ii) 48, (iii) 1973.*

The Greeks and numbers

We used to think that most of modern Western teaching came from Ancient Greece. They certainly made a huge contribution to literature and science but we now know that many of their ideas were borrowed from others. For example, instances of Pythagoras' Theorem were known in China, Egypt, and Babylon long before Pythagoras was alive. When it came to writing numbers the Greeks were not much more advanced than the Romans. They used two methods of writing numbers; one system called Herodian or Attic numbers, was very similar to Roman numerals. These are found in tax and accounting records.

Later, from the first century BC, they used the letters of their alphabet to write numbers. For example one was written as α (pronounced 'alpha'), two β ('beta'), three γ ('gamma'), and so on. In all they used 27 letters. Like Roman numerals their method did not lead to natural patterns and we can only wonder at the advances they did make with such a difficult system. Of course, much Greek mathematics was geometrical and did not use numbers.

7.2 Number bases and base ten

We have a highly developed system of writing numbers. It is called the *system of number bases* and in particular we usually use base ten. You may not have thought about the way we write numbers in columns with numerals but we hope this chapter gives you a greater sense of appreciation for this simple, elegant, and efficient system. You are all experts on this system already but we are going to tweak a few things in a moment so it will be worth us spending a few moments with a careful recap.

To start with we write the numbers between zero and nine using the *numerals*. These are;

Number	zero	one	two	three	four	five	six	seven	eight	nine
Numeral	0	1	2	3	4	5	6	7	8	9

When we reach ten we combine two of these numerals and write '10' which means 'one group of ten and no units'. Eleven is '11', 'one group of ten and one unit'. Of course we continue this in a systematic way. As we all know the numerals are combined in a string to write larger numbers. For example the three symbols 649 mean six groups of one hundred, four tens and nine units (or ones), or written out in a long and clumsy way

$$649 = 6 \times 100 + 4 \times 10 + 9 \times 1$$

or (and you might not be familiar with this notation here so hold tight...)

$$649 = 6 \times 10^2 + 4 \times 10^1 + 9.$$

You read 10^2 as 'ten to the power of two' (go on say it!). People often say 10^2 'ten squared' when the power is two. Can you work out why? In general when one number is written smaller and up to the right in this way we say 'to the power of' (go on say it!) just like '+' is pronounced 'add'. All 10^2 means is 10×10. Similarly 10^4 is short-hand for $10 \times 10 \times 10 \times 10$ (four tens in the product) and 10^1 is just 10 once. This is a real bit of mathematical notation and is very useful. Thus, in base ten the quantities represented by each column are the powers of ten:

	10^5	10^4	10^3	10^2	10^1	10^0
In base 10:	100000	10000	1000	100	10	1

In general, for two positive whole numbers a and b, a^b means

$$\underbrace{a \times a \times a \times \cdots \times a}_{b \text{ times}}.$$

For example $10^3 = 10 \times 10 \times 10$ and $2^5 = 2 \times 2 \times 2 \times 2 \times 2$. We will see a lot more of these kinds of calculations in Chapter 8.

Exercise. *Write out 7^4, 2^3, and 3^2 as a product. In each case calculate the value.*

Let's go back to the last expression for 649. That was $6 \times 10^2 + 4 \times 10^1 + 9$ which looks a lot more complicated! What we have achieved is to unwrap what the three symbols mean, namely 649 means six times one hundred plus four times ten and then nine ones. As we will see, understanding this *is* a big step forward.

Arabic numerals and modern notation

The exact origins of the system of number bases, in particular base ten, are hidden, but we know that from the middle of 2000 BC Indo-European tribes were making their way from what is now Afghanistan towards India. They used a language called Sanskrit and much of our knowledge of early mathematics comes from this time. In Sanskrit there are number words for 1–9, 10, 100, and further powers of 10, up to 10^{10}, so they definitely used a base ten system.

On its way to the West the Hindu method of writing numerals soon became known to the Arabs who had established a world empire from the seventh century. Many mathematical books from the Greeks and Hindus were translated in Baghdad, from where they later travelled to Western Europe. The Arabs played an essential part in transmitting the numeral system to us.

The best known individual was al-Khwarizmi. He was born about 780 in Bagdad, now part of Iraq, and died around 850. There is a dispute about the details of his life but what is known are details of the work he left us. He was the first Arab known to us to explain the Hindu system of numerals and two books he wrote, on 'Algebra' and 'Arithmetic', both survive in Latin translations.

The region of highest population density and fullest urban development in Medieval Europe was Italy – trade was good so they needed mathematics to keep track of business. Because of this they were interested in the place value system introduced by the Arab traders with whom they had links. Leonardo of Pisa, better known as Fibonacci, wrote his book '*Liber Abaci*' in 1202. In this he used the Hindu-Arabic system and this was a great success amongst merchants. From this time onwards this system spread throughout Europe.

Zero

Zero is a very mysterious number indeed. The symbol which is used to represent it is also used to show a space or gap. For example when we want to write down six hundred and three we have no 'tens'. We of course write this as 603 and use the numeral nought to show there is nothing in the tens column. Without such a symbol we might have to write this as 6 3, which is very similar to 63. Even worse, we couldn't tell if 6 represented six, sixty, six hundred, or something larger.

Zero really helps to make the point that numbers and the way we represent them are different. Read the following argument:

Person A: *What is zero?*
Person B: Well it's nothing.
Person A: *But it can't be 'nothing' because you called it zero and that is something.*
Person B: That's nonsense.
Person A: *And I'm confused!*

Confused? Well we are going to cut through this knot as follows. Take an empty box and stick a label 'empty' on the outside. The box is still empty. You see the label is not the thing itself. We use zero as a label to show the empty boxes or empty columns in our number system. This looks like sleight of hand but if we are going to avoid the confusion of our two friends above then it is a point we have to make. If you believe the explanation of the numeral nought, then we think you will avoid many of the pitfalls and confusions that have arisen in the past.

In Medieval times people argued long and hard about what zero was. They couldn't understand how by placing a symbol, which means nothing, after another symbol, the value increases by tenfold. For example putting '0' after '9' gives ninety. Remember that you are *never* allowed to divide by zero. It simply makes no mathematical sense. People were also very confused about this. Because of these mathematical and philosophical reasons people were very suspicious of using this system of writing down numbers. Because we are so familiar with base 10 and '0', that people were ever confused might come as a bit of a surprise. Having said that, please read the rest of the chapter with an open mind. We are about to introduce some new, and exciting ideas which *you* might otherwise treat with suspicion!

Another more practical reason why people were slow to accept the advantages of this system is that numbers written with numerals are easy to tamper with: Florence banned this system in 1299 precisely because you could falsify written records. It is just too simple to add a zero to an entry in an accounts book to make it look like there is more money than there should be. In some situations, such as writing the amount of money on a cheque, we still write out numbers with words. As mathematicians, of course, we clearly need something which is easier to calculate with.

In Europe people only started regularly writing down numbers in the way we do now after about 1550. People were very frightened of what was then a new number system that had been imported from the East and they were deeply suspicious. Up until that time they wrote numbers out long hand. Rather than writing 237 people actually wrote two hundred and thirty seven. This made maths much harder! Merchants also used Roman numbers but as we have seen for mathematicians this simply isn't as good. Negative numbers caused just as much confusion. In the end people settled for the symbol '$-$' for minus numbers. So -19 means 'minus nineteen'. People used to call negative numbers 'absurd' or 'fictional' amounts and often used them in a clumsy way.

7.3 Other number bases

There is no reason why we should group numbers together into tens and it hasn't always been this way. The Imperial system of weights and measures doesn't group things in tens. There are twelve inches in a foot for example, not ten.

Do you know:

- How many yards there are in a mile?
- How many pints there are in a gallon?

Does your answer to this question depend where you are in the world?

We call the number in a group the *base* of the system. As we will soon see we are not restricted to grouping things in tens. We can group things in many different ways.

Base 2: Binary

This is perhaps the simplest way of grouping numbers. Each column is only allowed to contain either a '1' or a '0'. As soon as we have more than one we take two from one column and put one into the column to the left. So now we are grouping things into twos not into tens. Lets start counting in binary and see how we get on.

	Base 10	Base 2	
one	1	1	
two	2	10	'One in the two's column and no units'
three	3	11	'One two + one unit'
four	4	100	'One four'
five	5	101	etc.
six	6	110	
seven	7	111	
eight	8	1000	
etc.			

Exercise. *Can you spot a pattern?*

Converting numbers to base two

There is an easy method to write a given number is base two. First write out a list of the *powers of two*. This means the list of numbers 1, 2, 2^2, 2^3, 2^4, 2^5, ..., etc., which means, 1, 2, 4, 8, 16, ... (in base 10 of course!)

The aim is to form a sum using the powers of two which adds up to our number.

To help with this we write the powers of two in the following grid:

	2^5	2^4	2^3	2^2	2^1	1
In base 10:	32	16	8	4	2	1

It makes life a lot easier so copy it out the first few times you do this until you can do it in your head. Now we use a simple and systematic procedure. Mathematicians call this an *algorithm* – a word derived from al-Khwarizmi's name.

1. Find all the powers of two less than or equal to your number.
2. Write a '1' under the largest and subtract this from your number.
3. If you have anything left think of this as a new number you are trying to write and go back to step 1.
4. When you have nothing left put zeros between the '1's in any empty columns and in all the columns to the right.

Although these four steps might look a little complicated all you need to do is practice them. Here is an example.

Example: Nineteen
Write down the powers of two:

2^5	2^4	2^3	2^2	2^1	1
32	16	8	4	2	1

Nineteen is less that $2^5 = 32$ and greater or equal to 2^4, $2^3, \ldots, 1$. The largest of these is of course 2^4 so we put a one under it

2^5	2^4	2^3	2^2	2^1	1
32	16	8	4	2	1
	1				

What have we got left? Nineteen minus sixteen (in base 10 this is $19 - 16$) is three. Let's repeat the process for three. When we do this we get:

2^5	2^4	2^3	2^2	2^1	1
32	16	8	4	2	1
	1			1	1

Next put zeros in empty columns:

2^5	2^4	2^3	2^2	2^1	1
32	16	8	4	2	1
	1	0	0	1	1

and so we write nineteen at 10011. Notice that at step 3 the number you are left with is always smaller. This means that you can't go on forever going round in circles and so the algorithm will eventually stop.

Remember, the aim is to write 19 using the powers of two. It is very easy and quick to check that we have indeed converted the number correctly. We have the binary number 10011 and of course this means (in base ten) $2^4 + 2^1 + 1 = 16 + 2 + 1 = 19$ and we have checked our working.

How do you know if '11' means eleven (grouping in tens) or three (in binary)? Of course you have to agree before you start the calculations and stick to one system throughout or otherwise you will get into an awful mess. In practice this doesn't cause a problem because people generally only ever use one system. Whenever you choose to use another system you should make this clear in your work.

It is *vital* as mathematicians and scientists to make clear which system we are using and stick rigidly to it when working. The crash of the Mars Climate Orbiter in 1999 was caused by two groups of scientists using different units: an embarrassing and expensive mistake.

The base two system is widely used. Modern computers, for example, work using binary. Inside a computer there are, crudely speaking, millions of tiny switches. These switches can either be on or off. If a switch is on we think of it as a one and if it is off we think of it as a zero. Thus we can 'turn on' a particular number by setting the switches to the corresponding binary number. In some very early computers the switches were actually turned on by hand but now the computer can manipulate its own switches to store numbers and do sums with them. The further exercises have a method of multiplying two numbers together which only works in base two and is very efficient. It is actually used in some computers to do multiplication.

Base 20: The Maya

The Mayan civilization of Central America, what is now southern Mexico, Guatemala, and Belize, flourished between about AD250 and 1000. From about 250 they had a highly developed culture, built monumental stone vaulted buildings and huge temple pyramids. After about AD1000 the population declined and the civilization finally came to an end after the invasion of the Spanish in 1521. In 1541 a catastrophe happened when the Franciscan monk Diego de Landa (1524–97) destroyed the Mayan library at Yucatan by setting fire to it. Later he tried to make amends by collecting as much material as he could but the damage was done and apart from the writing on buildings and temples, we have been left with only four Mayan books or *codexes*. Fortunately for us, much of what we do have is mathematical and astronomical.

0		10	
1		11	
2		12	
3		13	
4		14	
5		15	
6		16	
7		17	
8		18	
9		19	

Figure 7.4: Mayan numerals.

The Maya had a highly developed number system using, in part, base 20, including a zero. They used these numbers in conjunction with their calendar. In fact the Maya had two calendars operating at the same time. One, the sacred year, had 13 months of 20 days giving 260 days. The other, the vague year, has 18 months of 20 days and then a special month of 5 days to give a total of 365 days. The vague year was used for planting crops, etc., and the sacred year for religious festivals.

This might sound complicated but we have a number of systems cycling around each other, such as seven days in a week, days in each month, months in each year, lunar months, leap years, and so on.

The Maya built up the numerals for the numbers between 1 and 19 in a logical way using two symbols: a dot for a 1, a bar for a 5. They used a shell, the remains of a once-living body, to represent zero. The numerals are shown in Figure 7.4. They used these numerals to denote numbers in a positional system just like we do.

Base 60: The Babylonians

The Babylonian civilization existed about 4000 years ago in the Middle East in what is now Syria, Iraq, and Jordan. They lived in well-organized cities, had a highly developed culture and they have left some of the oldest written records. Many of these are mathematical, in the form of tax records and accounts. As was mentioned in Chapter 3 they were astronomers and studied the sun and the stars. To do this accurately they needed, and indeed developed, an efficient number system.

These people didn't write on paper in the way we do but used clay tablets and pressed a wedge-shaped stick into the clay. Next time you are in a museum ask if they have any tablets and see if you can find any mathematical ones on display. One simple mark was to make a wedge shape in the clay that looks something like this

The Babylonians used this mark to represent a one. Groups of marks could be made next to each other to give the numbers between 1 and 9 like this

The stick could also be turned sideways to give a similar shape, meaning ten, which looks like this

The sideways wedges (tens) and upright wedges (ones) were combined to give all the numbers between 1 and 59. As with the Maya, these groups of symbols acted as the numerals for their base 60 system.

Exercise. *What do the following patterns of wedges mean?*

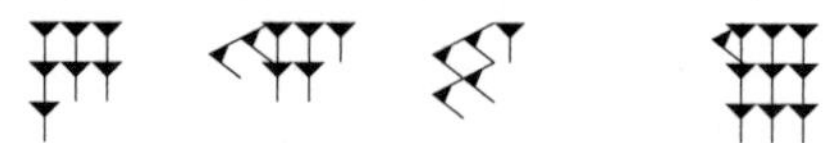

What happens when we get beyond 59? The remarkable thing about the Babylonians is that they developed a true system of number bases. Instead of inventing another symbol for 60 they used the same symbol for 1 and used it to show 'one group of sixty things'. For example, they wrote one thousand five hundred and fifty three (1553) as twenty five (25) groups of sixty and fifty three (53) units.

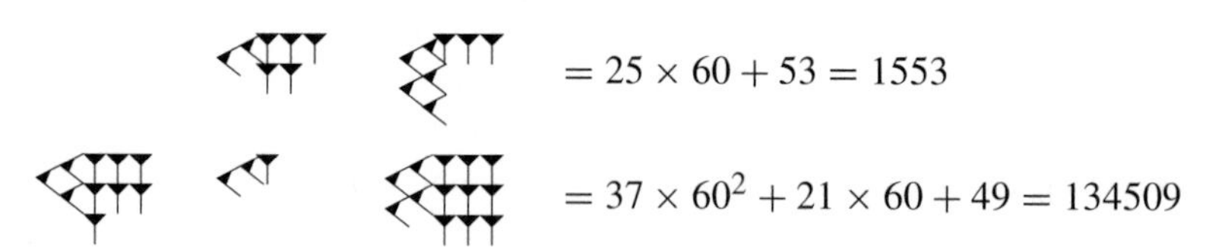

Similarly they wrote fractions in base 60 also. So the symbol for 1 would be 'one sixtieth' etc. Unfortunately the Babylonians didn't have a zero character or decimal point so

could be 15, 15×60, or $15/60$, etc. The way they got round this was to use other information from the problem. So, for example, if we are dividing pizza up equally

between friends and each person gets

it is most likely that each person gets 15/60'th of it, not 15 whole pizzas. (By the way, how many people eat the pizza?) It is unlikely that their lack of a zero caused much confusion as they left gaps and the problem provided a context. There is some evidence that later a symbol was used to show a gap or empty column but the use of this seems to have been limited.

We still use the remains of the Babylonian system today in the way we measure time and angles. There are sixty seconds in a minute and sixty minutes in an hour.

7.4 The counting board

A counting board is a simple extension of the tally system and it will allow us to explain the rules of arithmetic in terms of shuffling counters. Take any flat surface and divide it up with some straight lines. We will place counters onto this counting board. Systems like this were popular with merchants and traders because everyone could see what was going on. The counters could even be coins. Such a counting board is preserved in the Guildhall market in Bath. Known as 'The Nail' this waist-high marble pillar is where deals were made. Indeed, local lore has it that there is where the phrase 'cash on the nail' originates. So a counting board might look something like this with the board divided up into columns. The counters are represented by small circles:

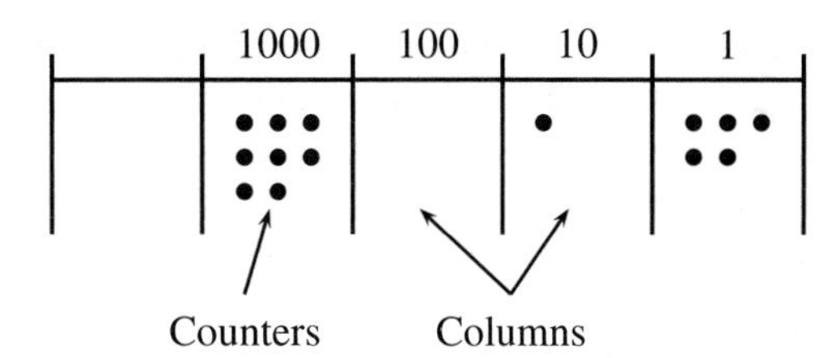

Transferring numbers to the counting board

To transfer a number to the counting board is very simple. We usually write numbers in base 10 and so we will do the same with our counting board. For example the number 8315 would appear on a counting board as

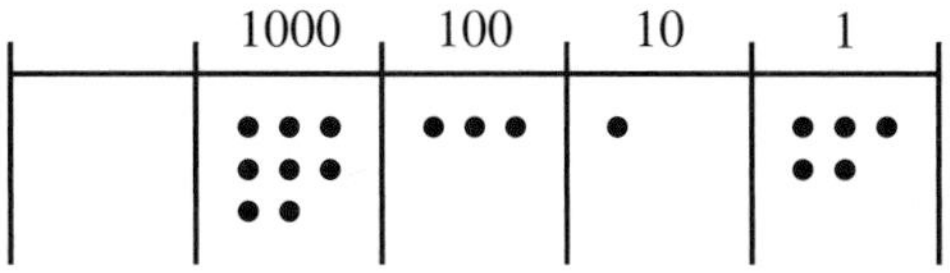

	1000	100	10	1
	••• ••• ••	•••	•	••• ••

We can think of arithmetic in terms of shuffling counters on the counting board. For example if we have ten or more counters in one column we can take ten of them off the board and replace them with one in the column to the left.

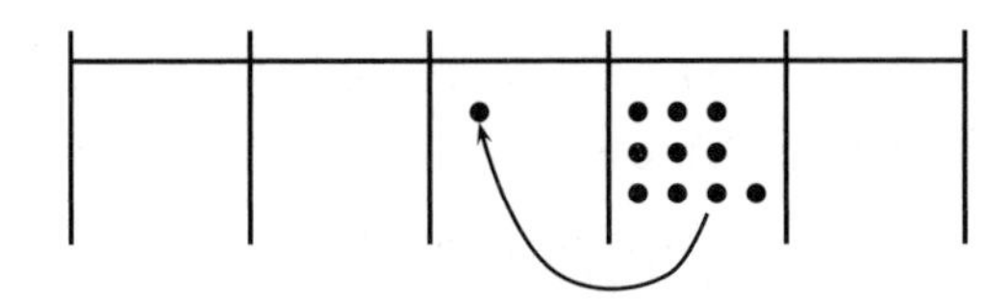

The advantage of this is that we can reduce arithmetic to a mechanical process.

7.5 Negative number bases

As we have seen there are many different systems of writing numbers. With the invention of digital computers the answers to previously abstract questions concerning number representations took on a new importance. How can we represent numbers inside our machines? What advantages do different methods have? How do we perform arithmetic operations? Which systems allow arithmetic to be performed efficiently? Since digital computers can easily distinguish an electrical 'on' state from an 'off' state the binary system was an obvious candidate. Indeed almost all computers use binary. But using the binary system we needed to introduce a new symbol '$-$' (minus) for negative integers. Does there exists a system in which both negative and positive numbers co-exists naturally? The answer to this is yes, and Donald Knuth, the famous computer scientist, gave one of the first written accounts of it in 1973. We choose to describe it here with base -2.

Earlier we saw that to represent a positive number in base two (binary) we write down the powers of two in columns starting at the right and working leftwards. Next we put a '1' in all the columns we *need* and then zeros in all the others remembering to drop any leading zeros. Our example was to show that nineteen was written as 10011. Next we will see how to write numbers using base -2. The processes of writing and manipulating numbers in base -2 will be, in essence, very similar to those of base 2.

To express nineteen in base -2
Write down the powers of minus two (remember that, for example $(-2)^3$ means $-2 \times -2 \times -2 = -8$).

	$(-2)^5$	$(-2)^4$	$(-2)^3$	$(-2)^2$	$(-2)^1$	1
In base 10:	-32	16	-8	4	-2	1

Add a one to the columns we *need*:

$(-2)^5$	$(-2)^4$	$(-2)^3$	$(-2)^2$	$(-2)^1$	1
-32	16	-8	4	-2	1
	1		1	1	1

That is to say, to get nineteen we need sixteen, four, minus two, and a one. Next add zeros so nineteen is written in base -2 as 10111. One question arises immediately,

Does this work?

To a mathematician, 'work' means:

(i) Can we write down every whole number (both positive and negative) using base -2?

(ii) Is there a number that can be written in more than one way?

Fortunately the answer is that 'yes', we can write each number down, and at least for whole numbers, 'no' there are no numbers that have two or more representations. This is a genuine mathematical theorem. We will *prove* part (i) later. This means we will supply a water-tight argument to convince you that it is true.

Exercise. *The following numbers are written in base (−2). Write them in base 10. (i) 100, (ii) 11001, (iii) 101000.*

Arithmetic in base -2

The next question one might well ask is "how do we add two numbers together"? Obviously converting them into binary or something else more familiar in which to do addition is cheating! Let's see if we can find some rules to add them directly. These rules will be mechanical, very much like moving round counters on a board. In fact we can use our counting board to illustrate this. By reducing the process of adding to a set of easy to remember rules we will make our life easier.

> **Awful warning!** The rules for adding, subtracting, carrying, and borrowing are different in base 2 and base -2.

We need to find new rules that allow us to calculate with our new way of writing numbers. It turns out that addition is very similar to normal binary addition but with very different 'carry' rules. In fact, the word 'carry' is a bit misleading and as we will see shortly it is more like a process of *absorption*. Notice that adjacent columns have opposite sign and so, assuming just for a moment that we are allowed to write a two in one column, the following are equal:

...	?	1	(2)	?	...

and

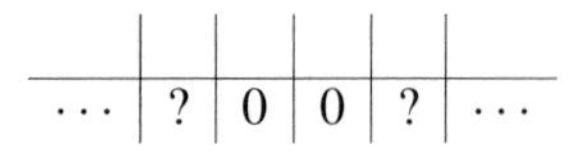

The two in the right-hand column cancels with the one in the column to the left. For example the right-hand column could be the 4's column and so the column to its left would be -8. Now, of course we aren't really allowed to place a 2 in a column but this slight abuse is no different from when we have to 'borrow' temporarily when we do normal subtraction. We will only have a two in a column in an intermediate step in a calculation.

The 2 and the 1 cancel each other out but of course we can do the reverse process to borrow if needed. In fact, we will need to do this when we add numbers together. Next we will show the method and then there will be some exercises for you to do. Please do these because practice will make negative number bases much easier to understand.

Adding numbers

To start, convert both numbers to base -2, and write them down one above the other as if you were about to do normal binary addition. To add them we start on the right and work progressively leftward, just as one would do in normal arithmetic. We only need to remember that the rules for 'carrying' digits are very different.

The carry rules

First note that

$$0 + 0 = 0$$

and

$$1 + 0 = 0 + 1 = 1$$

so in these simple cases there are no carries. The only case that requires some careful thought is the case $1 + 1$. We will explain this by examples. There are two cases we need to consider, $1 + (-1)$ and $1 + 1$. First

$$\begin{array}{rccc} & \cdots & 1 & 1 \\ + & \cdots & 0 & 1 \\ \hline & & 0 & 0 \end{array}$$

You can think of the two ones being *absorbed* by the one in the minus-two column.

Next consider the following example,

$$\begin{array}{rccc} & \cdots & 0 & 1 \\ + & \cdots & 0 & 1 \\ \hline \end{array}$$

Here we have nothing in the -2 column so we have to borrow and create two in the minus-two column and balance it with 1 in the fours column

$$\begin{array}{cccc} \cdots & 1 & (2) & 1 \\ + & \cdots & 0 & 1 \\ \hline \end{array} \quad \text{which gives} \quad \begin{array}{cccc} \cdots & 1 & (2) & 1 \\ + & \cdots & 0 & 1 \\ \hline & \cdots & 1 & 1 \quad 0 \end{array}$$

using the absorption rule on the two columns to the right. We can check this: 110 represents $4 - 2 + 0 = 2$ which is indeed $1 + 1$. Note the two is in brackets since we are not supposed to write a '2' anywhere as the only numerals are a 0 and a 1. If this is confusing think of it as being similar to normal subtraction when we borrow one and temporarily put ten in the column to the right while we perform the sum.

A worked example

Let's take a specific example and work through it using the counting board. The advantage of using a counting board rather than writing a 'sum' is that we can conveniently ignore the restrictions on the number of counters in a particular column in the intermediate steps in the calculation. Let's add seven to minus ten. First we need to write these in base -2. Seven is written as 11011 and minus ten as 1010 so that we must place counters onto the counting board as follows,

		16	-8	4	-2	1
		•	•		•	•
+			•		•	

Start on the right-hand column. This presents no problems as we only have one counter, so let's move this one counter down below the line.

		16	-8	4	-2	1
		•	•		•	
+			•		•	
						•

Now in the second column we have two counters so the rule says we must absorb one from the next column. But we can't do this yet as there isn't one in the third column. So we will add some counters to that we can achieve this. Compare this with *borrowing* in normal (base 10) subtraction.

		16	-8	4	-2	1
		•	•		•	
+			•		•	
			•	••		
						•

Now we take two counters from the second column and absorb them with one in the column to the left. This gives

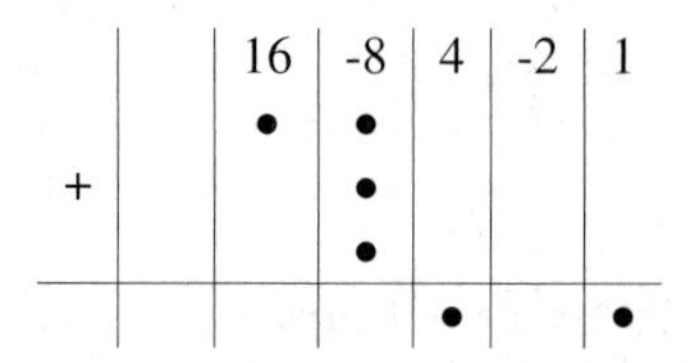

Lastly we take two counters from the fourth column and use these two to absorb the counter on the left. We are left with

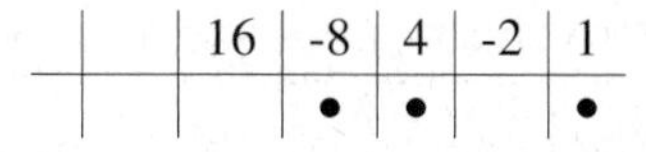

Thus on the counting board we get the result $11011 + 1010 = 1101$. Is this correct? Well, working backwards we see that 1101 means minus eight plus four plus one, i.e., minus three. Indeed seven minus ten is minus three (in base 10, $7 - 10 = -3$) and so it works! Try a few for yourself. Only by playing with these new rules and checking your answers will you be able to understand this. Also, don't be put off by a few mistakes at the beginning.

Yes, this does 'work'!

Now we will supply a water-tight argument as to why base (-2) works. Let's start at zero. We write this as 0 so that's a good start. Next let's assume we can write a number, n say. We know how to write one, '1' and minus one, '11'. Using the method above to add two numbers we can write $n + 1$ and $n - 1$. We can repeat this as many times as we need to, starting at zero to write any particular whole number we might choose. Voila!

7.6 Exercises

First session

1. What do the following Egyptian hieroglyphics mean?

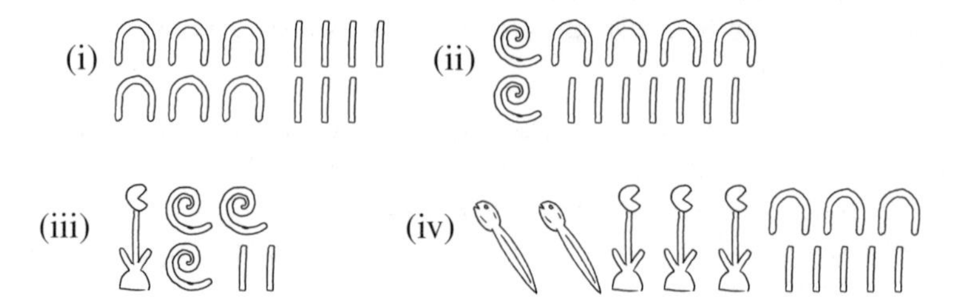

Figure 7.5: The Ancient Egyptian Army.

2. Write the following numbers using hieroglyphics: (i) seventeen, (ii) two hundred and thirty five, (iii) three million five hundred and seven thousand two hundred and four.
3. The Egyptian hieroglyphics run out for large numbers. Invent a few of your own for the numbers beyond a million.
4. Figure 7.5 is a copy of an ancient Egyptian hieroglyphic text which contains a lot of number symbols. Look at the picture symbols in the top two rows. They all face towards the right. Thus in this case we must read from *right* to *left*.

   The text indicates a block diagram showing the composition of fighting brigades of the Ancient Egyptian Army. The top row has two expressions which mean 'the fighters' and 'total force'. In the second row we see six expressions on top of six columns, from *right* to *left*

   (a) The first one is not clear exactly, but most likely it indicates 'infantry'.
   (b) The second one translates as 'fighting platoon no. 3'.
   (c) The third translates as 'marine fighting unit no. 5'.
   (d) The fourth, fifth, and sixth ones mean 'cavalry fighting brigade nos. 1, 2, and 3' respectively.

Your task is to

i. Translate the whole text including the numbers, keeping the grid layout in your answer.
ii. Find any relationships between the various numbers on the grid. Some are missing because the original text was damaged so you will have to make a guess at these.
iii. What do you think the total force is?

5. What do the following Babylonian numbers mean? (Assume there are no zeros.)

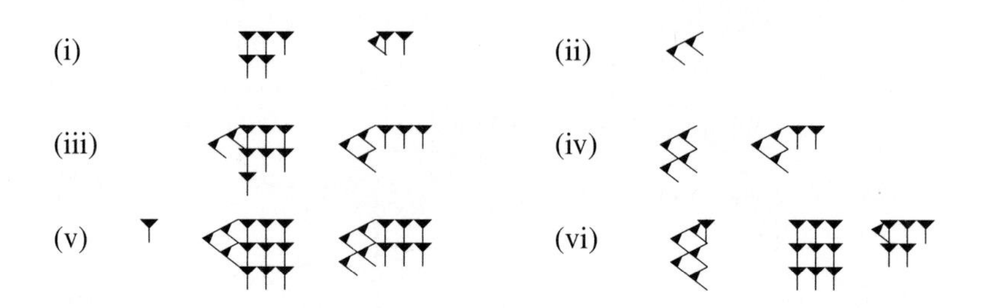

6. Find some counters. Dried beans or peas, pasta, etc., from the kitchen works well. Make a counting board with a blank piece of paper. Using a base 10 counting board – that is at most 10 counters in each column – work out the counting board moves to perform the following sums: (i) $7 + 12$, (ii) $35 + 8$, (iii) $92 + 909$, (iv) $1372 + 28\,249$.
7. Write out, in base ten, the powers of 2 up to 2^{20}. Write these as 2^0, 2^1, etc., and then calculate the values exactly. Now repeat this for the powers of minus two.
8. Write down the rules for addition and subtraction using a counting board as if you were explaining them to someone else. (Think of normal subtraction!)
9. Using the counting board rules perform the following calculations: (i) $12 - 6$, (ii) $45 - 23$, (iii) $107 - 29$.

Second session

1. Write the following numbers in base -2: (i) two, (ii) three, (iii) seven, (iv) minus five, (v) minus eight.
2. Using the counting board and rules for base -2 calculate

   (i) three + two
   (ii) seven − five
   (iii) seven − eight.

3. Explain the negative number base system to a friend. Set each other problems in base -2.
4. Take a number written in base -2, shift it left by adding a zero to the right-hand end. For example $1011 \mapsto 10110$. What does this transformation do mathematically?

5. What symbolic manipulation would allow you to multiply by -1? (Hint, use the answer to the above question!)
6. Carefully describe the process of addition in base -2. Give a step by step description.
7. What are the advantages of negative number bases? What are the disadvantages?

Field trips and projects

Most large towns have a museum. These contain a wealth of beautiful and interesting mathematical objects. For example, the Ashmolean Museum in Oxford is not a 'science museum', but it does have quite an impressive collection of mathematically interesting things if you know where to look! This includes Babylonian mathematical texts, an Egyptian slate ruler and school books, some Medieval sundials, and a Greek metrological relief. There are also some beautiful carved stone balls from Scotland in various geometric shapes and the oldest surviving Russian abacus. Can you find them all? A short visit to most museums and a chat with the staff, who are always pleased to help, should be all that is required to devise a 'mathematical treasure hunt'. We've tried these and they work well with students.

The British Museum, in London, has the Rind Papyrus and many examples of Babylonian cuneiform texts including a mathematical textbook.

7.7 Further problems

1. The following problems guide you through the problem of how to convert binary numbers into base -2.

   i. Write down (i) 2^{2n}, (ii) 2^{2n+1}, (iii) -2^{2n} and (iv) -2^{2n+1} in base -2.
   ii. Use the results from part (a) to convert positive numbers to base -2.
   iii. Find a way to convert a negative number to base -2 by first writing it in binary.

2. **Base -10**

   i. Write down the powers of (-10).
   ii. Express the following numbers in base -10: (i) seven, (ii) minus five, (iii) seventeen, (iv) ninety five.
   iii. Does base -10 'work', i.e., can you express every integer in base -10 and if so, is such a representation unique? Justify your answer.
   iv. Carefully describe the rules of addition in base -10.

3. **Multiplication in base 2.** We describe a simple way to multiply two numbers using base 2. This is the basis of the multiplication algorithms that many modern computers use. To start we need to write the two numbers in the product using base 2. To make the description easier to follow we will work through with the example of six times eleven which we write as 110×1011.

First, write one number along the top using a square grid to keep track of the columns. Write the other number down the side being careful to write the most significant digits, that is the first digits, *at the top*.

For our example you will end up with the following grid

	1	1	0				
1							
0							
1							
1							

Now copy out the first, horizontal number onto each line of the blank grid, shifting right at each stage so we get

	1	1	0				
1	1	1	0				
0	$\rightarrow$	1	1	0			
1		$\rightarrow$	1	1	0		
1			$\rightarrow$	1	1	0	

Next cross out all the lines with a zero in the left-hand columns and form a sum with the remaining rows. In our case the sum we get is

$$
\begin{array}{rccccc}
 & 1 & 1 & 0 & & \\
 & & & 1 & 1 & 0 \\
+ & & & & 1 & 1 \; 0 \\
\hline
\end{array}
$$

This sum is the answer we were looking for. If we use normal base 2 arithmetic to calculate this sum we get an answer 1000010. Checking this we see in base ten this is the number 66 and this is indeed six times eleven!

(a) Practise this by computing the following products (which are written in base ten): (i) 5×5, (ii) 5×6, (iii) 26×13.
(b) Practise this with products of your own. Check using your calculator.
(c) Why does this method work?

7.8 Answers

First session

1. (i) 67, (ii) 247, (iii) 1302, (iv) 203 035.

4. (a) The translated text is as follows

	Cavalry No. 3	Cavalry No. 2	Cavalry No. 1	Marine unit no. 5	Platoon no. 3	Infantry
		3		3	2	3
	3		2	2	1	2
	26	10		6	12	11
	30		13	12	12	21
		?	?	?	?	?
		3		4	2	3
		48	?	99	48	48
450	59	65	?	128	78	89

(b) The bottom row contains the total in the column above. The 450 in the column on the left is the total of the last row. This is like a modern computer spreadsheet! The missing numbers can be filled in to make the totals correct.
(c) The total force is 450.

5. (i) 312, (ii) 20, (iii) 1653, (iv) 2432, (v) 5986, (vi) 184155.
7. The powers of two are 1, 2, 4, 8, 16, 32, 64, 128, 256, 512, 1024, 2048, 4096, 8192, 16 384, 32 768, 65 536, 131 072, 262 144, 524 288, 1048 576. The powers of minus two are 1, −2, 4, −8, 16, −32, 64, −128, 256, −512, 1024, −2048, 4096, −8192, 16 384, −32 768, 65 536, −131 072, 262 144, −524 288, 1048 576.

Second session

1. (i) 110, (ii) 111, (iii) 11011, (iv) 1111, (v) 1000.

2. (i) $\begin{array}{rccc} & 1 & 1 & 1 \\ + & 1 & 1 & \\ \hline & 1 & 0 & 0 \end{array}$ (ii) $\begin{array}{rccccc} & 1 & 1 & 0 & 1 & 1 \\ + & & 1 & 1 & 1 & 1 \\ \hline & & & 1 & 1 & 0 \end{array}$ (iii) $\begin{array}{rccccc} & 1 & 1 & 0 & 1 & 1 \\ + & & 1 & 0 & 0 & 0 \\ \hline & & & & 1 & 1 \end{array}$

4. The shift $1011 \mapsto 10110$ corresponds to multiplication by −2, just as adding a zero to the end multiplies a number by 10 in base ten.
5. Shift the number as in question 4 above (i.e., multiply by −2) and then add the original number. For example 1011 is the number minus nine.

$$\begin{array}{rccccc} & 1 & 0 & 1 & 1 & 0 \\ + & & 1 & 0 & 1 & 1 \\ \hline & 1 & 1 & 0 & 0 & 1 \end{array}$$

The result of this calculation is (in base 10) $16 - 8 + 1$ which equals nine.

> To multiply by −1, left shift and add to itself.

Further problems

3. The sum at the end will be

(i)

$$\begin{array}{rccccccl} & 1 & 0 & 1 & & & \\ + & & & 1 & 0 & 1 & \\ \hline & 1 & 1 & 0 & 0 & 1 & (=25) \end{array}$$

(ii) There are two possible ways of writing this which give the two sums

$$\begin{array}{rccccc} & 1 & 0 & 1 & & \\ + & & 1 & 0 & 1 & 0 \\ \hline & 1 & 1 & 1 & 1 & 0 \end{array} \quad \text{or} \quad \begin{array}{rcccccl} & 1 & 1 & 0 & & & \\ + & & & 1 & 1 & 0 & \\ \hline & 1 & 1 & 1 & 1 & 0 & (=30) \end{array}$$

(iii) Again there are two ways to do this but it is always best to write the smaller number vertically giving

$$\begin{array}{rccccccccccl} & & 1 & 1 & 0 & 1 & 0 & & & & \\ & & & 1 & 1 & 0 & 1 & 0 & & & \\ + & & & & & 1 & 1 & 0 & 1 & 0 & \\ \hline & 1 & 0 & 1 & 0 & 1 & 0 & 0 & 1 & 0 & (=338) \end{array}$$

7.9 Mathematical notes

Fractions in base −2

Fractions can also be represented by negative number bases. To represent a fractional quantity we place the decimal point and then write negative powers to the right.

$(-2)^3$	$(-2)^2$	$(-2)^1$	$(-2)^0$	$(-2)^{-1}$	$(-2)^{-2}$	$(-2)^{-3}$	$(-2)^{-4}$
-8	4	-2	1	$-1/2$	$1/4$	$-1/8$	$1/16$

Using this a half (1/2) is written as 1.1 and a quarter is 0.01, etc.

Ordinal vs *cardinal numbers*

When we talked about numbers in this chapter what we meant was *quantity*. There is another way of thinking about numbers and that is in terms of *order*. The idea is to distinguish between having four apples as opposed to being fourth in the race. These ideas become much more important later in mathematics when we start to think about infinity but no description of ways of representing numbers would be complete without mentioning this very important difference.

One disadvantage of negative number bases is that we can no longer use the normal 'dictionary' ordering of strings of symbols to order numbers written in negative bases.

7.10 References

- Barrow, J. D. (1992). *Pi in the Sky.* Oxford University Press.
- McLeish, J. (1992). *Number.* Harper Collins.
  These two books contain much detailed and interesting information about the history and cultural significance of numbers.
- Knuth, D. E. (1969). *The Art of Computer Programming,* Volume 2: Seminumerical Algorithms, pp. 171 and exercises on pp. 176, 177, and 179. Addison-Wesley.
  This is the original article on negative number bases.
- Gardner, M. (1986). *Knotted Doughnuts and other Mathematical Entertainments.* Freeman.
  This excellent and accessible book, which contains a chapter on negative number bases, has many interesting topics that could easily be the subject of a masterclass.
- Wells, D. (1986). *The Penguin Dictionary of Curious and Interesting Numbers.* Penguin Books.
  This superb book contains a wealth of useful and fascinating information about numbers and their properties.

8
Doing the sums

8.1 Introduction

In Chapter 7 we looked at ways of writing whole numbers. The Hindu system of number bases, which we still use, turned out to be ideal and negative number bases are an interesting modern development. In this chapter we will introduce the logarithm. The discovery of the logarithm, by John Napier, was perhaps the most important advance in practical calculation methods after the introduction of the Hindu, decimal number system. The logarithm, like many things in mathematics, turns out to have a much deeper significance and greater use than simply helping us calculate.

In the second half of this chapter we go on to describe a practical implementation of logarithms: the slide rule. The slide rule is also an example of what is known as an *analogue computer*. Most modern computers are digital and they store and manipulate information as strings of ones and zeros. We saw binary arithmetic in Chapter 7. Some computers, in particular early computers, were not digital and used other technologies, such as flowing liquids or mechanical links. The importance of these things are now becoming forgotten, particularly with the rise of the electronic calculator. We ask you to do as many of the exercises as you can with only paper and pencil. Of course, your calculator is useful to help you to check your answers.

Many of the events that take place in this chapter occurred in the early seventeenth century. This was a time of great activity in the arts, science, literature, and world travel. It is the period after the Dark and Middle Ages when European scholars rediscovered the classical writings of the Ancient Romans and Greeks which had been preserved and translated by Arab scholars. People began to make voyages overseas, for example the Pilgrim Fathers began to travel to the Americas around 1620. This is the time of Shakespeare and Milton.

It was also a time of religious and political turmoil. King Henry VIII of England, who died in 1547, had split the Christian Church in England. In 1588 the Spanish Armada was defeated by Sir Walter Raleigh and Guy Fawkes was thwarted in his attempt with the Gunpowder Plot to destroy the King and Parliament in 1605. Later during this period the English Civil War occurred, which started in 1642, and towards the end we see the Great Plague (1665), and the Fire of London the following year. It is not perhaps surprising then to find great advances in mathematics and science. But let's start right at the beginning of this exciting and important story.

8.2 Before logarithms

People have always tried to find ways of reducing the effort required to perform calculations. The Hindu system of writing numbers was one such advance but multiplication and particularly division were still difficult and fraught with errors. In 1544 Michael Sifel wrote his book *Arithmetica Integra*. In this he wrote down the following two series of numbers one above the other.

$$\left.\begin{array}{cccccccccc} 0 & 1 & 2 & 3 & 4 & 5 & 6 & 7 & 8 & \dots \\ 1 & 2 & 4 & 8 & 16 & 32 & 64 & 128 & 256 & \dots \end{array}\right\} \quad (8.1)$$

To calculate the next in the entry in the top line we add one, in the bottom line we multiply by two. In modern notation the bottom series is simply 2^n. Sifel pointed out that addition in the upper series corresponds to multiplication in the lower series. For example take 3 and 5 from the top line. Written below these are 8 and 32 respectively. Thus,

$$\begin{array}{ccccc} 3 & + & 5 & = & 8 \\ \downarrow & & \downarrow & & \downarrow \\ 8 & \times & 32 & = & 256 \end{array}$$

Exercise. *Use (8.1) to calculate* 8×16.

Sifel lacked the modern notation we have and so would have been unable to calculate 8×32 by writing

$$8 \times 32 = 2^3 \times 2^5 = 2^{3+5} = 2^8 = 256$$

and thus transforming his multiplication to a simple addition using the rule that

$$2^a \times 2^b = 2^{a+b}.$$

Sifel made some remarks about these series of numbers but it was John Napier who made a major breakthrough and this led to the idea of a logarithm.

> Logarithms can be used to make calculations easier by transforming a multiplication to an addition and a division into a subtraction.

Exercise. *Write out a series, similar to (8.1) but using powers of* 10 *rather than* 2.

8.3 The life of John Napier

John Napier was born in 1550 at Merchiston Castle in Scotland. On the death of his father he became the eighteenth Lord of Merchiston. Napier studied at St Andrews University and travelled widely round Europe. He published *A Plaine Discovery of the Whole Revelation of Saint John* in 1593 which he regarded as his most significant work and it passed through several editions in English, French, German, and Dutch. It was a new mathematically inspired interpretation of the book of Revelation from the Bible and in it he claims the Pope was the 'Antichrist'. He also predicted that the world was due to end in 1786 but fortunately this particular calculation was incorrect!

In July 1594, Napier entered a strange contract with a baron, Robert Logan of Restalrig, who had just been outlawed. In the contract he agreed to try and discover some kind of treasure that was rumoured to be hidden in Logan's dwelling-place, Fast Castle. Napier was to receive a third of the treasure when found and the contract appears to show that Napier believed in magic.

In 1596 he published details of inventions designed to defend his own castle. One of these included a parabolic mirror with which he could set fire to enemies at a great distance by concentrating the rays of the sun. An almost identical story is connected with the destruction of the Roman fleet, by the great Greek mathematician Archimedes (287–212 BC), at Syracuse. It is uncertain if Napier actually carried out the experiments but with the amount of sunlight in Scotland in winter the device cannot have been an effective deterrent to a would-be attacker.

What we remember Napier best for is the small book, first published in 1614, entitled *Mirifici Logarithmorum Canonis Descriptio* (Description of the Wonderful Canon of Logarithms). This book was a set of tables and a short description of how to use them. An explanation of the theory and the method of calculation was published after his death in 1619 as *Mirifici Logarithmorum Canonis Constructio* (Construction of the Wonderful Canon of Logarithms).

John Napier's geometric ideas

Napier's definition was based on the continuous movement of two points, since this was the only way he had of considering continuously varying quantities. Not only did Napier lack the calculus to bring rigour to his ideas but, again, he lacked modern notation. All his arguments are described, at length, in plain language. In fact Napier's logarithm, as described in terms of moving particles, differs slightly from the modern one which we are about to describe. The full story of the calculus and in particular John Napier's contributions to this can be found in the references at the end.

Henry Briggs and the common logarithm

In 1615 Henry Briggs, an English professor of mathematics, visited Napier in Scotland. Briggs was very impressed and suggested that Napier's logarithms could be substantially improved and he immediately started to compute a set of improved logarithms which had much more useful computational properties. These are what we still use today and describe here, but actually calculating such tables is much more difficult. Briggs also translated Napier's *Constructio* and then published a set of incomplete tables after many years of paper and pencil calculations.

The effect of logarithms on seventeenth century science

Late sixteenth century developments in astronomy and navigation created a demand for ever more lengthy and difficult trigonometric calculations. Before logarithms calculations were tedious. Some algebraic tricks helped, such as the *quarter squares* formula

$$4AB = (A+B)^2 - (A-B)^2$$

and the formula

$$2\sin(A)\sin(B) = \cos(A-B) - \cos(A+B).$$

Trigonometrical and other sets of tables were also compiled and used, such as those of Georg Joachim Rheticus (1514–1576) whose calculations for a collection of 15-place trigonometrical tables were published in 1613. Rheticus is perhaps the greatest human computer ever to have lived. His tables included natural sines for every $1/360$ degree to 15 places of decimals, all calculated by hand! The effort required to calculate such tables provided the motivation for Charles Babbage to design and build his mechanical computers at the end of the nineteenth century. His difference engine no. 2 has been built at the National Museum of Science and Technology in London and is well worth a visit.

The uses of logarithms were obvious to the whole scientific community and Napier's ideas caught on quickly. When Johann Kepler (1571–1630) received a copy of Napier's tables he used them to assist with the calculations which led to the formulation of his laws of planetary motion. Without the aid of logarithms these calculations would have taken many years. In fact Kepler published a letter addressed to Napier as the dedication to his *Ephemeris* or star catalogue of 1620. This congratulated Napier for his invention and describes the benefit they would have on astronomy.

Kepler's laws gave Newton evidence which was crucial to support his theory of universal gravitation. It can genuinely be said that Napier laid the foundations on which the whole of the development of the calculus and hence most of modern science and technology was subsequently built. Napier clearly understood the importance of his own work, despite his death only four years after publication in 1617. In his dedication to Prince Charles he writes,

Most Noble Prince,

Seeing there is neither study, nor any kind of learning that does more arouse and stir up generous and heroical wits to excellent and eminent affairs: and contrariwise that does more deceive and keep down drunken and dull minds, than the Mathematics. It is no marvel that learned and magnanimous Princes in all former ages have taken great delight in them and that unskillful and slothful men have always pursued them with most cruel hatred, as utter enemies to their ignorance and sluggishness.

He continues in the preface to the reader:

Seeing there is nothing (right well beloved Students in the Mathematics) that is so troublesome to Mathematical practice, nor that doth more lest and hinder Calculators, than the Multiplications, Divisions, square and cubical Extractions of great numbers, which besides the tedious expense of time, are for the most part subject to many slippery errors. I began therefore to consider in my mind, by what certain and ready Act I might remove those hindrances.

The invention of logarithms proved the usefulness of the Hindu system of numerals and number bases we examined in Chapter 7 which was still not generally accepted even in Napier's day. Given that writing whole numbers was confusing to many people then, writing fractions as decimal fractions was even more of a problem! Evidence for this comes from Oughtred's book, *The Circles of Proportion and the Horizontal Instrument* (1624) which we will talk more of later. When explaining why his inventions were not widely used he wrote that "*The former of which two scruples ariseth from the ignorance of the true nature and manner of Decimal Fraction.*" He then gives a detailed explanation of decimal notation and its use. Rather than using the decimal point, as we do now, Oughtred, like Briggs, underlined the fractional part of the number. He gives the example, $37\underline{06}$ rather than $37\frac{6}{100}$. It was John Napier who invented the decimal point, as we now know it, and Napier's systematic use of decimal notation was widely responsible for its adoption in the seventeenth century. We hope this gives you some idea of the significance of his discoveries.

8.4 What are logarithms?

Unfortunately we will define logarithms by looking at functions and their inverses. This means our definition will be, in some senses, back to front which makes it a little tricky. So, before we define the logarithm we will look closely at *functions* and their *inverses*.

This section is quite involved and introduces some sophisticated mathematical ideas. We hope you enjoy working through them, but if you find it too difficult feel free to skip forward to the next section which tells you how to *use* logarithms, and come back to this section later.

Imagine you have a positive number x, say. You could, if you so wished, multiply x by itself to give x-squared or x^2. This process is a particular example of what mathematicians often call a *function* and we often write this as

$$f : x \mapsto x^2 \quad \text{or} \quad f(x) := x^2.$$

We read this as 'x maps to x-squared'.

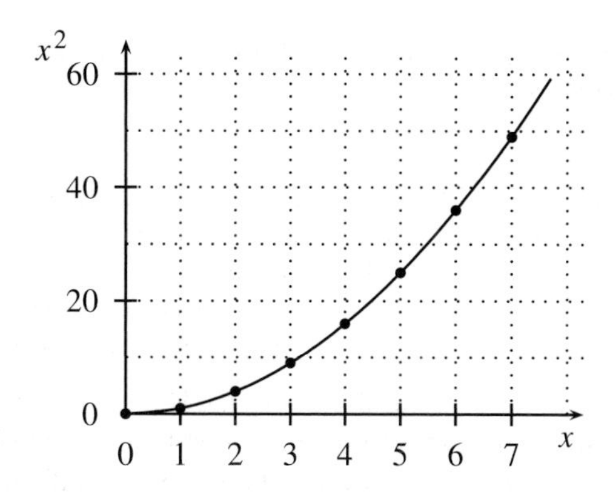

Figure 8.1: The graph of x^2.

Strictly speaking to define a function you need:

(i) a collection, D, (called the *domain*) of things (such as a set of numbers) that you are going to supply to the function,

(ii) a rule, f, (such as $x \mapsto x^2$), and

(iii) another collection, R, (called the *range*) of things that you could end up with at the end.

All you require is that you can apply the rule to every x in D and you get back something in R.

Exercise. *Is it possible to apply the rule $x \mapsto x^2$ to every number x?*

This is all very well and a bit abstract, but we more often think of a function in terms of a picture that represents it. This is called the *graph*. We've plotted part of the graph of $x \mapsto x^2$ in Figure 8.1.

Another separate, but also interesting, question we can ask is: given $y > 0$, can we find a unique number x such that $y = x^2$? In the case of our function $x \mapsto x^2$ when $x > 0$ we can always find such a number. By changing the rule or even set D for the domain this might not be the case.

Exercise. *Given $y = 9$ what $x > 0$ satisfies $x^2 = 9$?*

This is the opposite way around and is known as the *inverse* of f. Of course after a little thought we realize that the inverse of this function is the square root, written $\sqrt{x}$. Another example is given in Figure 8.2 where we find $\sqrt{30}$ by using the graph in Figure 8.1.

Inspired by the notation

$$x^a \times x^b = x^{a+b}$$

we often write $\sqrt{x} = x^{1/2}$.

Exercise. *Show that $x^{1/2} \times x^{1/2} = x$.*

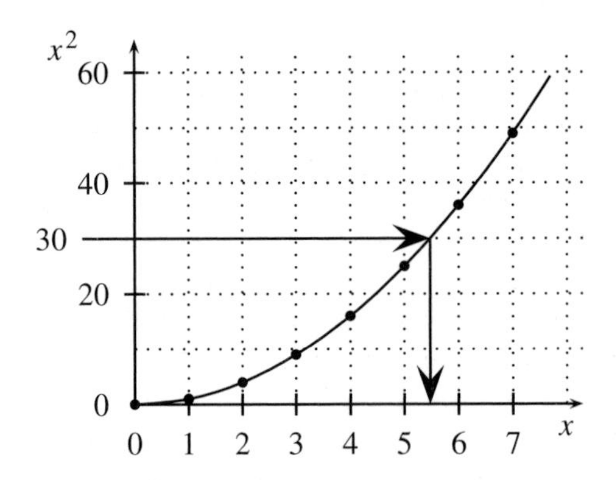

Figure 8.2: Finding the inverse from a graph.

Similarly $x^{1/3}$ is the cube root of x, and $x^{1/q}$ is the q-th root of x. All these roots exist but if we gave you a number, x say, actually calculating them would be quite another matter! For now we will just use a calculator but do keep in it in mind that this is a recent luxury that previous generations of mathematicians had to do without.

The logarithm – at last!

In Section 8.2 we looked at raising 2 to the power of x. We also asked you to raise 10 to the power x. Now we will perform this process in reverse to calculate the logarithm of a positive number y. To find the logarithm of y what you have to do is to try and write y as 10^x for some other number x.

The problem:

Given $y > 0$, find a number x such that $y = 10^x$.

We then say that the *logarithm* of y is x and write this as

$$\log_{10}(y) = x.$$

We call 10 the *base* of the logarithms. Although we only define the logarithm for positive numbers $y > 0$ there is nothing special about taking 10 as the base. If we took 2 as a base and tried to solve $64 = 2^x$ then comparing this with Michael Sifel's series of numbers, (8.1), at the beginning of this chapter we see that $\log_2(64) = 6$ (since $2^6 = 64$).

We chose to use 10 as the base because your calculator almost certainly has this in-built. It doesn't matter that we make this restriction but, as when working with number bases, we have to make sure we agree before we start.

The first question that we should ask is 'does this make sense?' Given a number $y > 0$ can we calculate its logarithm? If so, how would we calculate it?

Now, if y is a power of 10 then calculating the logarithm is easy: just compare it with the series we asked you to write down in Section 8.2. For example $\log_{10}(1000) = 3$, but what happens if $y = 450$ say? This number doesn't appear in the series so we need to solve the equation $450 = 10^x$. Since our series is increasing and $100 < 450 < 1000$ we might think, with some justification, that if $10^x = 45$ then $2 < x < 3$. But thinking about this for a moment might throw up an obvious question. 10^2 means 10×10. What does 10^x mean when x is not a whole number?

The logarithm on a graph

How can we possibly get an idea of what all this looks like? One way is to plot the graph of the points which are easy to calculate. These are,

x	0	1	2	3	4	5...
y	1	10	100	1000	10 000	100 000...

Next, if we can find $\sqrt{10}$ we can plot $10^{1/2}$, $10^{3/2}$, $10^{5/2}$, $10^{7/2}$, ..., etc. Using cube roots and p-th roots we can plot points to whatever accuracy we might need. In fact, we can calculate $x^{p/q}$ for any whole numbers p and q. It turns out that this will enable us to draw a *smooth curve* to connect all the points together, just as we did for $x \mapsto x^2$. This is shown in Figure 8.3.

Now we can read off the graph the logarithm of 450. But we have to be careful. In this case we need to look for 450 on the y-axis, that is the *vertical* axis. Back to front indeed!

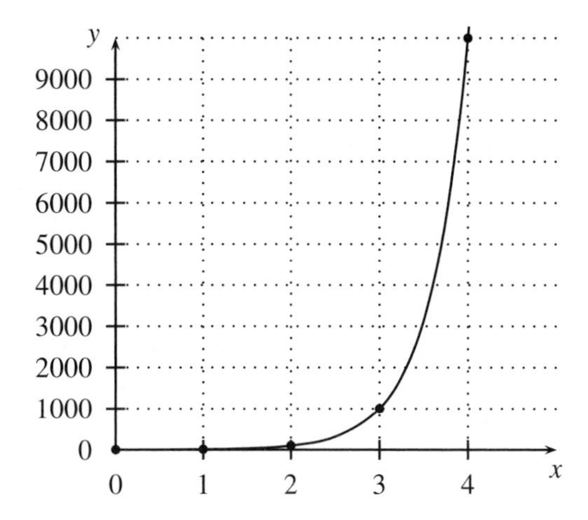

Figure 8.3: The graph of 10^x.

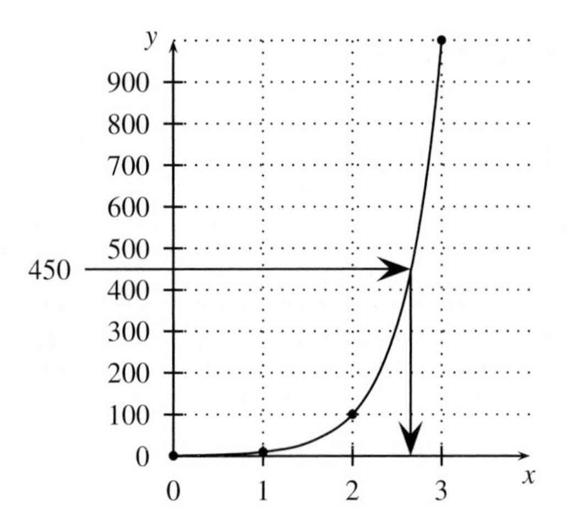

Figure 8.4: The logarithm as an inverse function.

Looking carefully at the horizontal axis we see the line cuts it at a point close to 2.6. Therefore we conclude that the logarithm of 450 (with base 10) is 2.6 to within the accuracy of the graph. What we are claiming is that $10^{2.6} \approx 450$. In fact, using a calculator we find $\log_{10}(450) = 2.653\,212\,51\ldots$ so our graphical estimate is not at all bad.

The process of drawing a graph for points we know and then using it for points we don't know is so important it has a special name: *interpolation*. In practice people don't carry round graphs, they use tables instead. We are going to concentrate on using tables rather than justifying their theory. This will, unfortunately, leave some mathematical gaps, some of which are addressed in the mathematical notes.

8.5 Using tables of logarithms

Up until the 1970s, when someone needed the answer to calculations that were complicated, or that required an accurate answer, a set of tables were used. Using tables to aid calculations isn't hard and we will show you how do so using Table 8.1 of logarithms, to base 10. In this section we will assume that when 'logarithm' is mentioned it is in base 10. We use 10 as a base because it is the basis of our usual number system. This makes calculations easier.

Logarithms of numbers between 1 and 10
To find the logarithm of a number $1 \leq x \leq 10$ we use Table 8.1. For example if $x = 6.7$ look along the 6 row and down the 0.7 column and get the answer $\log_{10}(6.7) = 0.826$.

Exercise. *What is the logarithm of* 3.9*?*

Table 8.1: Logarithms to base 10.

	0.0	0.1	0.2	0.3	0.4	0.5	0.6	0.7	0.8	0.9
1	0.000	0.041	0.079	0.114	0.146	0.176	0.204	0.230	0.255	0.279
2	0.301	0.322	0.342	0.362	0.380	0.398	0.415	0.431	0.447	0.462
3	0.477	0.491	0.505	0.519	0.531	0.544	0.556	0.568	0.580	0.591
4	0.602	0.613	0.623	0.633	0.643	0.653	0.663	0.672	0.681	0.690
5	0.699	0.708	0.716	0.724	0.732	0.740	0.748	0.756	0.763	0.771
6	0.778	0.785	0.792	0.799	0.806	0.813	0.820	0.826	0.833	0.839
7	0.845	0.851	0.857	0.863	0.869	0.875	0.881	0.886	0.892	0.898
8	0.903	0.908	0.914	0.919	0.924	0.929	0.934	0.940	0.944	0.949
9	0.954	0.959	0.964	0.968	0.973	0.978	0.982	0.987	0.991	0.996
10	1.000									

To go in the other direction, given a logarithm of 0.556 we look for this entry in the table. We find this entry appears in the 3 row and 0.6 column. Thus $\log_{10}(3.6) = 0.556$ or remembering that the logarithm of y is the solution of $y = 10^x$, $10^{0.556} = 3.6$.

Exercise. *What number has logarithm* 0.929*?*

What happens if the logarithm doesn't appear in the table? In this case you have to find the two numbers closest and estimate. Imagine a smooth curve, like the one we drew in Figure 8.3, and use this picture in your mind. For example the number 0.866 does not appear in the table. How then do we calculate $10^{0.866}$? First we notice that $10^{0.863} = 7.3$ and that $10^{0.869} = 7.4$. Now 0.866 is half way between 0.863 and 0.869 but the graph is *not a straight line* so $10^{0.866}$ will not be exactly half way between 7.3 and 7.4. But we are going to approximate $10^{0.866}$ as 7.35 because the table doesn't supply any more information. In fact $10^{0.866} = 7.345$ to three decimal places so 7.35 isn't too bad.

When using tables there will be round-off errors. These errors are the difference between the values in the table and the true values. For example the table gives $\log_{10}(6.7) = 0.826$ whereas the true value is closer to 0.826 078 48 .... These errors are one of the disadvantages of tables over a calculator. However, understanding they exist and experience of allowing for them will give you a deeper understanding of what the numbers mean. In fact, calculators also produce a roundoff error but this is *very* small, around 10^{-10}, so we usually don't notice it. In the above example the error was about 5/1000 or 1/200. This might be acceptable but if not, you need to find a set of tables which give more accurate values such as those in Table 8.2 given at the end of this chapter.

Exercise. *If you pace out a distance of* 1 km *and you measure this to within* 1/200 *how large is your error?*

Using the tables to multiply

One (perhaps the most) important feature of logarithms lies in their importance in *practical* computations. Here it is.

The idea:

To multiply two numbers add their logarithms.

By this we mean

$$\log_{10}(a \times b) = \log_{10}(a) + \log_{10}(b).$$

Why is this the case? It all stems from the simple rule

$$10^x \times 10^y = 10^{x+y}.$$

Let's do an easy example and use the tables to calculate 2×3. Using the table $\log_{10}(2) = 0.301$ and $\log_{10}(3) = 0.477$. The rules says that

$$\log_{10}(2 \times 3) = \log_{10}(2) + \log_{10}(3) = 0.301 + 0.477 = 0.778.$$

Sure enough if you look up $\log_{10}(6)$ the tables will give you the answer 0.778.

Another example, 2×4.

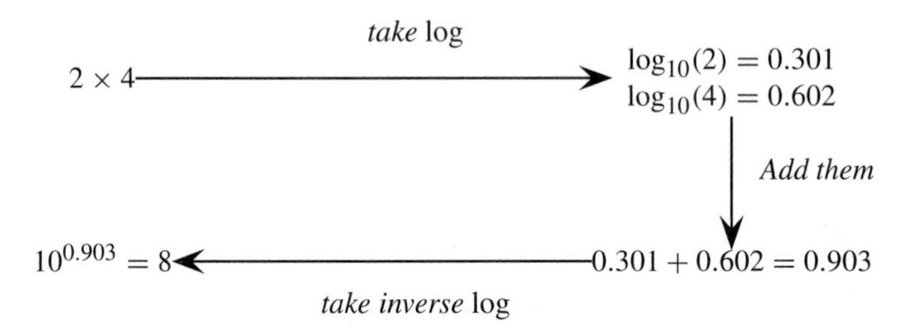

These examples were easy so now let's try an example that does not involve whole numbers. What about 2.1×3.8? From the table $\log_{10}(2.1) = 0.322$ and $\log_{10}(3.8) = 0.580$. Adding these gives $\log_{10}(2.1 \times 3.8) = 0.902$. To get the answer we need to calculate $10^{0.902}$ which the table tells us is just less than 8. Hence the rule of logarithms tells us that 2.1×3.8 is just less that 8. In fact $2.1 \times 3.8 = 7.98$ so to within the accuracy of the tables we get the correct answer.

Exercise. *Show that* $\log_{10}(10b) = 1 + b$.

The above exercise shows why it is useful to express logarithms to base 10 as the common process of multiplication by 10 transforms to the simple process of the addition of 1.

Logarithms of other numbers

What happens if you want to find the logarithm of a number which is greater than 10? For example what is $\log_{10}(37)$? First remember that we are trying to solve the equation $10^x = 37$ for some number x. Now from the tables we know that $10^{0.568} = 3.7$. Multiply both sides by 10 to get $10^{1.568} = 10^{1+0.568} = 10 \times 10^{0.568} = 37$.

> To find the logarithm of a number first multiply or divide by 10 until the number lies between 1 and 10. Use the table to find the logarithm of this part and then add or subtract the number of times you multiplied by 10.

For example to find $\log_{10}(790)$ first find $\log_{10}(7.9) = 0.898$ and then remember that $790 = 100 \times 10^{0.898} = 10^{2.898}$. Thus $\log_{10}(790) = 2.898$.

As another example, to find $\log_{10}(0.002)$ first find $\log_{10}(2) = 0.301$ and then $0.002 = 10^{0.301} \times 10^{-3} = 10^{0.301-3} = 10^{-2.699}$. Thus $\log_{10}(0.002) = -2.699$.

The reverse is similar. If you want to calculate $10^{4.959}$ then note that $10^{4.959} = 10^4 \times 10^{0.959}$ and from the tables this is equal to $10^4 \times 9.1 = 91000$.

There are plenty more examples for you to try in the exercises.

Division

To understand how to use logarithms to help with division we need to remember the simple rule

$$\frac{10^x}{10^y} = 10^x \times 10^{-y} = 10^{x-y}.$$

Using this we get the rule for division

$$\log_{10}(a/b) = \log_{10}(a) - \log_{10}(b).$$

For example let's calculate $36/12$. First, as before we write $36 = 10 \times 3.6$ and $12 = 10 \times 1.2$ so that

$$\frac{36}{12} = \frac{10 \times 3.6}{10 \times 1.2} = \frac{3.6}{1.2}.$$

We do this because the tables only show the logarithms of the numbers between 1 and 10. Next we look up the values $\log_{10}(3.6) = 0.556$ and $\log_{10}(1.2) = 0.079$. The rule for division gives us that $\log_{10}(3.6/1.2) = 0.556 - 0.079 = 0.477$. To get the answer we seek we have to find $10^{0.477}$ and then we will know that $36/12 = 10^{0.477} = 3$.

8.6 The slide rule

The idea of a slide rule is as simple as it is beautiful. We can 'add' numbers very simply by taking two rulers and placing them side by side. By sliding one past the other we can perform simple sums. The diagram in Figure 8.5 illustrates this idea.

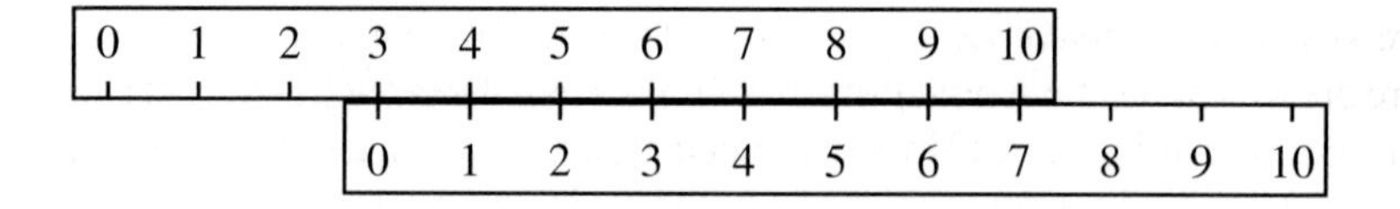

Figure 8.5: Adding numbers with a slide rule.

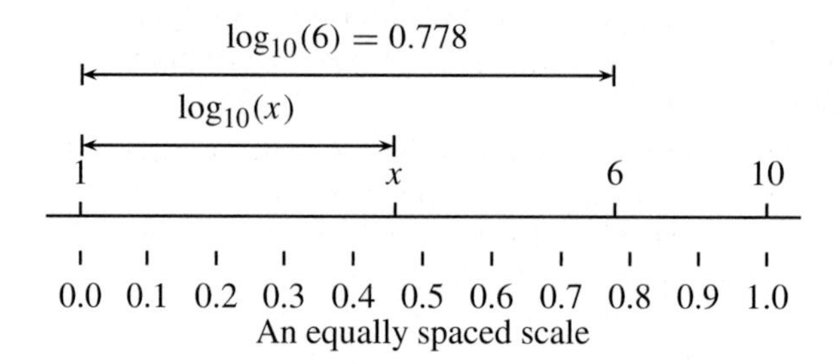

Figure 8.6: Drawing a logarithmic scale.

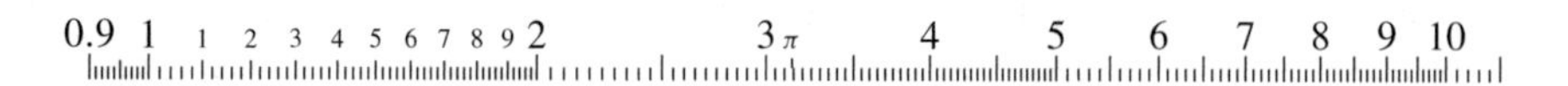

Figure 8.7: A completed logarithmic scale.

The numbers on the bottom are three less than the numbers on the top and using these we can 'add three' easily.

As we have seen, logarithms allow us to convert a multiplication into addition. If we can make a ruler with a logarithmic scale we could use this to add logarithms and this in turn will correspond to multiplication of the numbers. How would we do this? We first draw a line and mark two points on it. These will correspond to the numbers 1 and 10 and we call the length of this line 1. To mark a point on the scale, 6 for example, we measure the logarithm of 6 along the line. This is illustrated in Figure 8.6.

A completed logarithmic scale is show in Figure 8.7. Unlike a normal ruler the distance between 1 and 2 is greater than 2 and 3. This is because the graph of the logarithm function is not a straight line. Mathematicians say that the logarithm is a *non-linear* function.

Reading a logarithmic scale

The first thing to learn is how to read a logarithmic scale. This does take a bit of practice and is best mastered by quietly and carefully examining the scales.

The first thing to notice is that the distance between the marks is uneven and that there are more marks between 1 and 2 than between 9 and 10. This is because there is more space on the scale. Between 1 and 2 there are 49 marks but between 9 and 10 there are only nine. A common source of mistakes is forgetting this! It also means that the scale is more precise between 1 and 2 than between 9 and 10.

Using a slide rule to multiply

Remember that to multiply two numbers all you need to do is add their logarithms. In just the same way we used the rulers in Figure 8.5 to add numbers, we can multiply two numbers by using two logarithmic rulers. For example let's take as an example the simple product 2×3.

Take two logarithmic rulers. We call them C and D because these letters are traditionally used on slide rules for the scales shown here. Place them side by side so that the 1 on C is next to the 2 on the D. Now look on C for the 3. The answer to the product 2×3 will be opposite on the D scale, as shown in Figure 8.8.

We have added the logarithms of 2 (on the D scale) to the logarithm of 3 (on the C scale). You can see this more clearly in Figure 8.9.

Notice that *without moving the ruler* we can do lots of other products. For example we can read 2×4 and 2×5 off the slide rule without touching the ruler. This is useful and quicker than using a calculator!

What happens when we get to 2×6? Looking below the 6 on the C scale we notice we have run out of markings on the D scale. If you ever run out of scale like this you

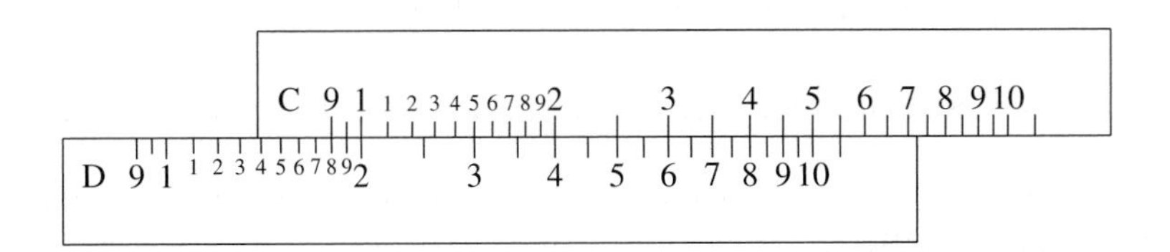

Figure 8.8: Setting two logarithmic rulers to perform 2×3.

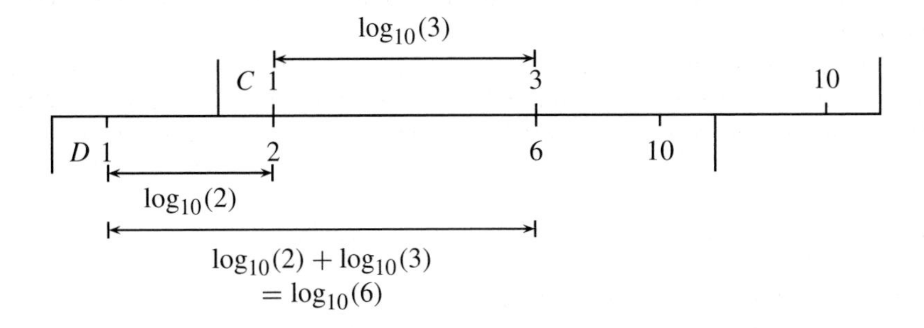

Figure 8.9: The principle of a slide rule.

need to *divide by ten*. This corresponds to moving the whole of the C ruler to the left so that the 10 on the C is where the 1 currently is. The rulers should now look like this

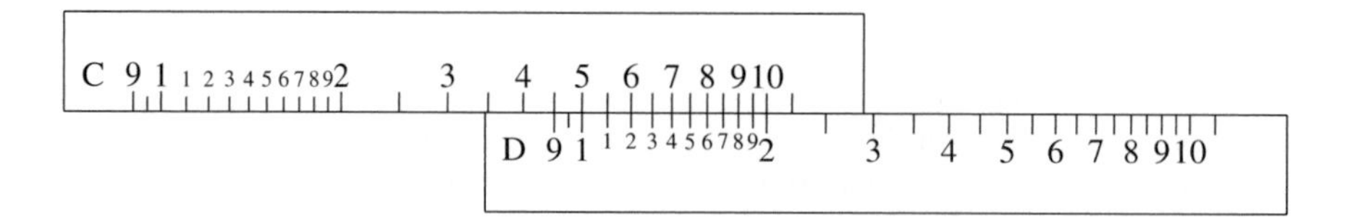

Now look at the rulers. The 6 on C is now above 1.2 on D but we *divided by ten* so the answer must be 12 which is, of course, 2×6. Similarly we can read off 2×7, 2×8, etc.

> Remember: when using a slide rule we have to keep track of the decimal point ourselves.

We certainly don't think of this as a disadvantage. By keeping track of the decimal point when doing calculations we constantly have to know what we are doing and so we are less likely to make a mistake. Having to keep track of the decimal point in this way is similar to what we need to do when using logarithm tables.

The circular slide rule

The second solution to the problem of running out of scale on a slide rule is to use a circular one. This has another advantage: by wrapping a scale round a circle we get, on a circle of given diameter, more scale than on a straight rule of that given length. In fact the first ever slide rule was circular and a circular logarithmic scale is shown in Figure 8.10.

We can use this by placing one scale inside another. For example the two scales in Figure 8.11 are set up in such a way that we can use them to multiply by π. Thus, we can use Figure 8.11 to find the circumference of a circle given its diameter. If the inside scale corresponds to the diameter then the outside will show the circumference. Take a circle of diameter 1. Then the circumference will be π as shown on the outside. The inside of 2 is just below 6.3 on the outside and the inside 5 is next to 1.57 on the outside giving a circumference of 15.7 because we have been round once and have to remember to multiply by ten.

The advantage is that we don't run out of scale in the same way that we do on a linear rule. Of course, we still need to keep track of the decimal point.

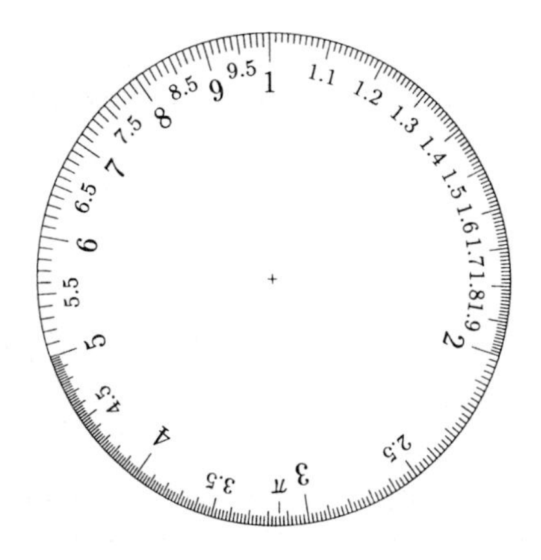

Figure 8.10: A circular logarithmic scale.

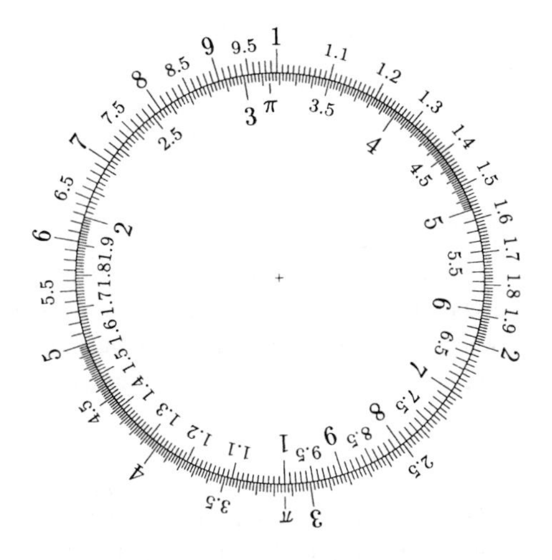

Figure 8.11: Circular rule set to multiply by π.

8.7 Who invented the slide rule?

Edmund Gunter (1581–1626)

Edmund Gunter's most important book entitled *Description and Use of the Sector*, was first published in English in 1623. This has been described as *'the most important work on the science of navigation to be published in the seventeenth century.'* A sector is a mathematical instrument consisting of two hinged arms on which there are engraved scales which can be used to help with calculations. This is not a slide-rule; the single scale is used in conjunction with a pair of compasses. In practice the points of the compass tend to damage the scales which reduces the accuracy of the instrument. What makes Gunter's sector special is that it is the first mathematical instrument to be inscribed with a logarithmic scale to help solve numerical problems.

William Oughtred (1574–1660)

William Oughtred was a clergyman and keen mathematician. He is believed to have introduced the × symbol for multiplication in his book *Clavis Mathematicae* (Key to Mathematics), written about 1628 and published in London in 1631. This was a very important maths textbook at the time. Newton read it and was influenced by it for example.

He is now generally thought to be the inventer of the slide rule. Circular rules are described in a book with the title *The Circles of Proportion and the Horizontal Instrument*, London, 1632. There is good evidence to suggest that Oughtred had made his invention a number of years earlier and failed to publish it. The first instrument he describes is the *Circle of Proportion* itself. He says in the description of how to use it that *Numbers are multiplied by addition of their logarithms; and they are divided by subtraction of their logarithms*. This is exactly how we described the use of logarithms and thus the principle behind the slide rule.

Richard Delamain (1600–1644)

Richard Delamain, a teacher of mathematics, was originally William Oughtred's student. He taught Charles I for example and the first published account of a slide rule as we would recognize it was given in his book *Grammelogia*. This was a 30-page pamphlet which described a circular slide rule. The instrument consisted of two metal disks held together at the centre with a small pin.

Both Oughtred and Delamain argued over who invented the circular slide rule in their various books over the years. Their friends also got involved and things became heated. Delamain was a big slide rule fan, arguing that mechanical aids helped people understand how to calculate. Oughtred was not. He didn't publish his notes earlier because he thought using slide rules only reduced mathematicians to 'doers of tricks' and each accused the other of stealing the invention. Certainly Oughtred's circle of proportion is more detailed and versatile than Delamain's mathematical ring. It is generally believed that Oughtred and Delamain were independent inventors. Oughtred got there first but Delamain was the first to publish.

8.8 Exercises

First session

1. Use Table 8.1 to find the logarithm of the following numbers:
   (i) 2.5, (ii) 4.0, (iii) 3.2, (iv) 42, (v) 97, (vi) 0.2.

2. (a) Use Table 8.1 to calculate (i) $10^{0.146}$, (ii) $10^{0.633}$, (iii) $10^{0.929}$.
   (b) Use Table 8.1 to estimate (i) $10^{0.61}$, (ii) $10^{0.525}$, (iii) $10^{0.989}$.
3. Perform the following calculations using the logarithms given in Table 8.1. Question 1 might help for some of them!
   (i) 2.5×4.0, (ii) 2×7, (iii) 7.2×3.2, (iv) 42×97, (v) 0.2×15.
4. *The area of a circle.* Given the logarithm of π is 0.497 (3 d.p.) find the area of a circle of the following radii:
   (i) 3.7, (ii) 5.2, (iii) 9.1.
5. *Square roots.* We can write $\sqrt{y} = y^{\frac{1}{2}}$ since

$$\sqrt{y}\sqrt{y} = y^{\frac{1}{2}}y^{\frac{1}{2}} = y^{\frac{1}{2}+\frac{1}{2}} = y.$$

This gives us a method of working out $\sqrt{y}$ using logarithms. If $x = \log_{10}(y)$ then $\frac{1}{2}x = \log_{10}(\sqrt{y})$.

As a worked example we use this method to find $\sqrt{9}$ (even though we already know $\sqrt{9} = \pm 3$). First we look in Table 8.1 and find the value of $\log_{10}(9)$ is 0.954. $0.5 \times 0.954 = 0.477$ and if we look up $10^{0.477}$ in Table 8.1 we find $10^{0.477} = 3$. Thus we conclude that $\sqrt{9} = \pm 3$ (remembering that a positive number has *two* square roots).

Use the tables and this method to find the square roots of the following numbers. (i) 4, (ii) 16, (iii) 5.3.

Adapt this method to find cube roots using the fact that $\sqrt[3]{y} = y^{1/3}$. This requires a little thought. Use your method to find $\sqrt[3]{27}$ and $\sqrt[3]{216}$.

Second session

1. Find a dictionary and look up *accurate* and *precise*. What, if anything, is the difference? Can you find examples of simple mathematical statements that are (i) accurate but not precise, (ii) precise but not accurate, (iii) accurate and precise.
2. Look at the following logarithmic scale. Read the points marked (a)–(e).

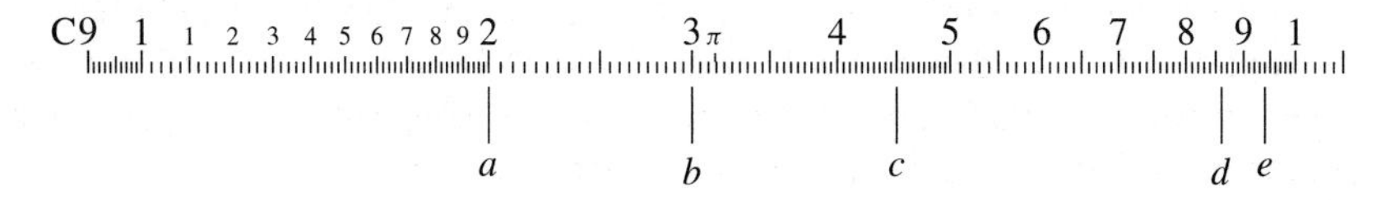

3. Look at the following logarithmic scale. Read the points marked (a)–(e).

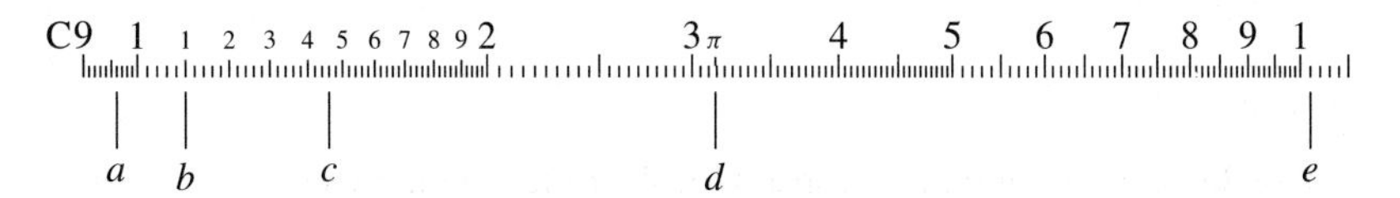

4. Look at the scale in Figure 8.7. Locate the following points on the scale; (a) 5, (b) 9, (c) 6.3, (d) 5.5, (e) 1.1, (f) 1.74, (g) 0.92.
5. Use the following slide rule to perform the following calculations. (a) 2.5×2, (b) 2.5×3, (c) $2.5 \times \pi$.

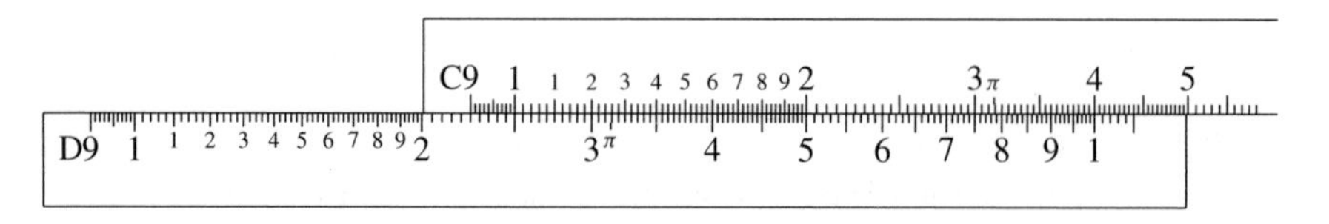

6. Look at the scale on the circular rule in Figure 8.11. Use it to calculate (a) $2 \times \pi$, (b) $3.5 \times \pi$, (c) $7 \times \pi$.
7. *Make your own slide rule.* Using the ideas from Figure 8.9 carefully make a slide rule of your own. Be warned this is quite time consuming! Think about which numbers to inscribe. Test out your design.
8. If you have a slide rule of your own examine it carefully. Write a list of all the scales and what mathematical functions they represent. Give examples of calculations using them.

Field trips and projects

As mentioned, the National Museum of Science and Technology in London have built Charles Babbage's Difference Engine no. 2. They also have a wonderful collection of slide rules and other mathematical and calculating instruments. This museum makes an excellent visit. For more information see their Web site
`http://www.nmsi.ac.uk/`

It might seem bizarre to recommend second-hand bookshops and car-boot sales but this is precisely the kind of place in which you will find an old slide rule or book of log tables. Although some are collectors items it is still quite easy to find a basic slide rule, such as that used by generations of university students, in good condition and that doesn't cost more than a few pounds. Sometimes parents or friends will have one so ask around. If you do manage to purchase or borrow one please look after it. Make sure that the rule slides easily and the parts are intact. If you need to clean it be very careful, it is easy to damage the scales. Remember this is a precision scientific instrument so don't use it to draw straight lines with pens that will damage the scales or mark the rule.

As the proud owner of a slide rule, work out what all the scales show and try and find ways of using them. Much time was devoted to laying out combinations of scales to make trigonometrical calculations simple and accurate. As Obi-Wan Kenobi said about the Light Sabre in Star Wars IV, a slide rule is 'an elegant weapon for a more civilized time'.

Further problems

1. *Converting to other logarithmic bases.* We have seen two kinds of logarithms, logarithms to base 10 and logarithms to the base 2. Recall that in each case,

$$\text{Given } y > 0,\ x = \log_{10}(y) \text{ is the solution to } y = 10^x$$

and

$$\text{Given } y > 0,\ x = \log_2(y) \text{ is the solution to } y = 2^x.$$

What is the connection between $\log_{10}$ and $\log_2$? To answer this let's start by solving the equation

$$2 = 10^\alpha$$

for some α. How do we do this? Easy! $\alpha = \log_{10}(2)$ and from Table 8.1, $\alpha = 0.301$. If we need to find $\log_2(y)$ we need to solve $y = 2^x$ which is the same as $y = (10^\alpha)^x = 10^{\alpha x}$. So given $\alpha x = \log_{10}(y)$ we can divide both sides by α to get that $x = \log_2(y) = \log_{10}(y)/\alpha$, i.e.,

$$\log_2(y) = \frac{\log_{10}(y)}{\log_{10}(2)}.$$

Let's try to use this formula to calculate $\log_2(64)$ which we already know, from Michael Sifel's series (8.1), is 6. From Table 8.1 we have $\log_{10}(2) = 0.301$ and $\log_{10}(64) = 1.806$. Using your calculator to check, we have $1.806/0.301 = 6$!

> We only need tables of logarithms to base 10 to calculate *all* other logarithms to *all* other bases $a > 0$ using the formula
>
> $$\log_a(y) = \frac{\log_{10}(y)}{\log_{10}(a)}.$$

Use this, Table 8.1, and if you need, to your calculator, to find
(i) $\log_2(1024)$, (ii) $\log_7(343)$, (iii) $\log_{16}(0.5)$.

2. *Natural logarithms.* In the above question we saw how to calculate logarithms with different bases. Is there any reason why the base has to be a whole number? No, there isn't and a very important logarithm, called the *natural logarithm*, uses a base e which is approximately

$$e = 2.718\,281\,828\,459\,045\,235\,360\,287\,47\ldots$$

Like π this number pops up all over the place in mathematics and you will certainly encounter it again. We have already seen the link between π and e in Chapter 6.

Sir Isaac Newton (1643–1727) discovered the following formula for calculating e,

$$e = 1 + \frac{1}{1} + \frac{1}{1 \times 2} + \frac{1}{1 \times 2 \times 3} + \frac{1}{1 \times 2 \times 3 \times 4} + \frac{1}{1 \times 2 \times 3 \times 4 \times 5} + \cdots$$

Remembering the short-hand $n! = 1 \times 2 \times \cdots \times (n-1) \times n$, this can be written as

$$e = 1 + \frac{1}{1!} + \frac{1}{2!} + \frac{1}{3!} + \frac{1}{4!} + \frac{1}{5!} + \cdots$$

which we think is particularly beautiful. Since, for large n, the value of $1/n!$ becomes very small indeed, we can approximate e accurately by forgetting about the terms for large n. If we do this we end up with a *finite sum*, which we can, in principle at least, calculate.

For each n, we will define the *partial sums* σ_n by the formula

$$\sigma_n := 1 + \frac{1}{1!} + \frac{1}{2!} + \frac{1}{3!} + \cdots + \frac{1}{n!}$$

so that $\sigma_1 = 2$, $\sigma_3 = 5/2$ and $\sigma_3 = 1 + 1 + 1/2 + 1/6 = 16/6$. Notice that to calculate σ_n we only need to work out σ_{n-1} and the last term by using the formula

$$\sigma_n = \sigma_{n-1} + \frac{1}{n!}.$$

Use this to complete the following table:

σ_n	Fraction	Decimal
σ_1	2	2
σ_2	5/2	2.5
σ_3	16/6	2.666...
σ_4	65/24	
σ_5		
σ_6		
σ_7		
σ_8		
σ_9		

Compare your decimal values with the value given above.

3. *A time line.* Take a blank sheet of paper and work through the last two chapters to create a time line showing the major civilizations and events. Mark on known dates and the people who figure prominently. You could illustrate each civilization with examples of the numbers they used and mark on other important dates you know.

8.9 Answers

First session

1. (i) 0.398, (ii) 0.602, (iii) 0.505 (iv) 1.623, (v) 1.987, (vi) $0.301 - 1 = -0.699$.
2. (a) (i) 1.4, (ii) 4.3, (iii) 8.5.
   (b) (i) $10^{0.61} \approx 4.07$, (ii) $10^{0.525} \approx 3.35$, (iii) $10^{0.989} \approx 9.75$.
3. (i) $0.398 + 0.602 = 1.0$ and $10^1 = 10$ so $2.5 \times 4.0 = 10$.
   (ii) $0.301 + 0.845 = 1.146$ and $10^{1.146} = 14$.
   (iii) $0.857 + 0.505 = 1.362$ and $10^{1.362} = 23$.
   (iv) $1.623 + 1.987 = 3.610$ and $10^{3.610} = 10^3 \times 10^{0.610} \approx 4070$.
   (v) $-0.699 + 1.176 = 0.477$ and $10^{0.477} = 3.0$.
1. (i) $\log_{10}(3.7) = 0.568$. Using the formula $A = \pi r^2$ for the area of the circle we calculate $0.497 + 2 \times 0.568 = 1.633$. The area will be $10^{1.633} = 10 \times 4.3 = 43$. The area is 43 to within the accuracy of the tables.
   (ii) Similarly $\log_{10}(5.2) = 0.716$ and $0.497 + 2 \times 0.716 = 1.929$ which gives the area as $10^{1.929} = 10 \times 8.5 = 85$.
   (iii) Lastly $\log_{10}(9.1) = 0.959$ and $0.497 + 2 \times 0.959 = 2.415$ which gives the area as $10^{2.415} = 100 \times 2.6 = 260$.

Second session

1. First the definitions. *Accurate*: Lacking error, conforming exactly with the truth. *Precise*: exact in statement.
   These are quite different. Examples are (i) 'π is about 3'. This is accurate but not precise. (ii) $\pi = -2.718\,281\,828\ldots$. This is precise (exact in statement) but not accurate since it is not true! (iii) $\pi = 3.1415927\ldots$ is both accurate and precise. Of course there are degrees of accuracy and precision here which are entirely subjective but we hope this is illustrative.
2. (a) 2.0, (b) 3.0, (c) 4.5, (d) 8.6, (e) 9.4.
3. (a) 0.96, (b) 1.1, (c) 1.46, (d) $\pi = 3.14$, (e) 10.2.
4. 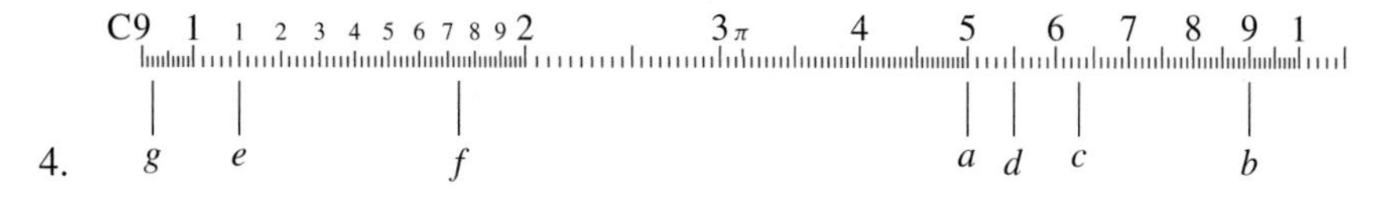

5. (a) 5, (b) 7.5, (c) 7.85.

Further problems

1. (i) $\log_2(1024) = \frac{2.708}{0.301} \approx 9$,
   (ii) $\log_7(343) = \frac{2.531}{0.845} \approx 3$,
   (iii) $\log_{16}(0.5) = \frac{0.699-1}{1.202} \approx -0.25$.

2. Dots above digits indicate the recurring part of the decimal, for example $0.\dot{3}$ means $0.333\,33\ldots$.

σ_n	Fraction	Decimal
σ_1	2	2
σ_2	5/2	2.5
σ_3	16/6	$2.\dot{6}$
σ_4	65/24	$2.708\dot{3}$
σ_5	163/60	$2.71\dot{6}$
σ_6	1957/720	$2.718\,0\dot{5}$
σ_7	685/252	$2.718\,\dot{2}53\,9\dot{6}$
σ_8	109601/40320	$2.718\,278\,\dot{7}69\,841\,\dot{2}$
σ_9	98641/36288	$2.718\,281\,525\,5\cdots$

The difference $e - \sigma_9 \approx 3 \times 10^{-7}$. The approximation $\sigma_7 = 685/252$ is within about 3×10^{-5} of the true value and, as a fraction, is fairly easy to remember. In terms of measuring a length, 10^{-5} corresponds to 1 cm in 1 km so σ_7 could easily be used in practice without causing too much error.

8.10 Mathematical notes

We have avoided many interesting but technical questions here and it might well be fair to say that we have glossed over much that is important. One of these is the fact that, as strange as it might appear at first, *not all* numbers can be written as fractions.

What exactly do we mean by the statement 'the number x cannot be written as a fraction'? We mean that no matter how we choose the whole numbers p and q we can't find two which satisfy

$$x = \frac{p}{q}.$$

For example we have already defined the number $\log_{10}(2)$ as being the solution to the equation

$$10^x = 2.$$

Now imagine that we could write $x = \log_{10}(2)$ as a fraction. There would be two whole numbers p and q so that

$$10^{\frac{p}{q}} = 2$$

or, equivalently,

$$10^p = 2^q.$$

Let's think about 10^p. Since p is a whole number, 10^p written in base 10 will end in a zero. Can 2^q ever end in a zero? No, and there are two ways to see this.

First think about numbers that end in a zero. You can divide them by 10 and so you can divide them by 5. Because q is a whole number 2^q is just a string of 2's multiplied together and so five, which is a prime number, will not divide 2 (another prime). This might not convince the more sceptical of you and so we now supply a more convincing argument by using some group theory. Perhaps you should look back at Chapter 2 to remind yourself, and in particular look at the clock groups in the problems.

We are only interested in the *last digit* of a particular product. A few moments thoughts should convince you that the last digit of a product of two numbers a and b will be the last digit of the product of the last digits of a and b. Put another way if a has $0 \leq \alpha \leq 9$ as its last digit and b has β as its last digit then $a \times b$ will have the last digit of $\alpha \times \beta$ as its last digit. Remember we are looking to see if 0 is ever a last digit of 2^q.

If we start looking at last digits we see that

q	2^q	last digit
1	2	2
2	4	4
3	8	8
4	16	6
5	32	2
6	64	4
etc.		

Looking at this series we notice that all the numbers are even and the only even number that does *not* appear is 0! To convince you this is true look at the following table. In fact this is the same as the group table for the clock group with four elements. It shows the last digit of the product of the two numbers α and β.

	6	2	4	8
6	6	2	4	8
2	2	4	8	6
4	4	8	6	2
8	8	6	8	4

In particular 0 never appears in the table so we can't generate 0 as the last digit of 2^q for any whole number q. Thus $10^p \neq 2^q$ for all whole numbers p and q. This proves that not all numbers can be written as a fraction. There are lots of other examples of numbers you are already familiar with that cannot be written as a fraction. For example neither π, e, or $\sqrt{2}$ can be written as fractions although we don't prove it here.

The ancient Greeks first noticed that not all numbers could be written as fractions by thinking about the length of the diagonal of a unit square, which has length $\sqrt{2}$. By a careful argument they proved that writing this length as a fraction was impossible. This radically altered their view of numbers and for the discoverers, the Pythagoreans, it was one of their innermost secrets. In fact legend has it that one of their brotherhood was executed by being thrown from a ship and drowned for leaking it to the outside world! Their very short and beautiful proof is a standard part of any university mathematics course.

The logarithm has not been defined for negative numbers in this chapter. Suffice it to say that just as the Greeks had to expand their idea of what a number was to include things, like $\sqrt{2}$, that cannot be written as a fraction, we would have to expand our idea of numbers to include $\log_{10}(-1)$. Leonard Euler, one of the greatest mathematicians, called these new numbers *complex* which is a shame! They are not complex at all. Quite the reverse: they simplify things immensely and modern mathematics would be quite impossible without them.

8.11 References

- Edwards, C. H. (1979). *The Historical Development of the Calculus,* Chapter 6. Springer Verlag.
  This book is the definitive history of the calculus and Chapter 6 is devoted to John Napier's logarithm and tables.
- Kelles, L. M., Kern W. F., and Bland, J. R. (1943). *The Log–Log Duplex Slide Rule: A Manual.* Keuffell & Esser Co.
  There are many slide rule manuals although most are out of print and hard to get hold of. This book is reasonably common but almost every slide rule came with its own instruction book and there are many others.
- Hopp, P. M. (1999). *Slide Rules: Their History, Models and Makers.* Astragal.
  There have been very few books published on slide rules. This book is aimed at the serious collector and contains a comprehensive catalogue of slide rules.

Web sites

- `http://www-groups.dcs.st-and.ac.uk/~history/index.html`
  History of mathematics at St Andrew's University.

8.12 Cut out and keep

In the next pages are some 'cut-out and keep' slide rules. Photocopy them, cut them out carefully, and glue them onto stiff card. Use them to help with the exercises if you need to. Caution – many photocopiers distort the images and so you will need to check the accuracy of your new 'cut-out and keep' slide rule with a few simple calculations.

A 9 1 2 3π 4 5 6 7 8 9 1 2 3 4 5 6 7 8 9 1

D 9 1 1 2 3 4 5 6 7 8 9 2 3 π 4 5 6 7 8 9 1

B 9 1 2 3π 4 5 6 7 8 9 1 2 3 4 5 6 7 8 9 1

C 9 1 1 2 3 4 5 6 7 8 9 2 3 π 4 5 6 7 8 9 1

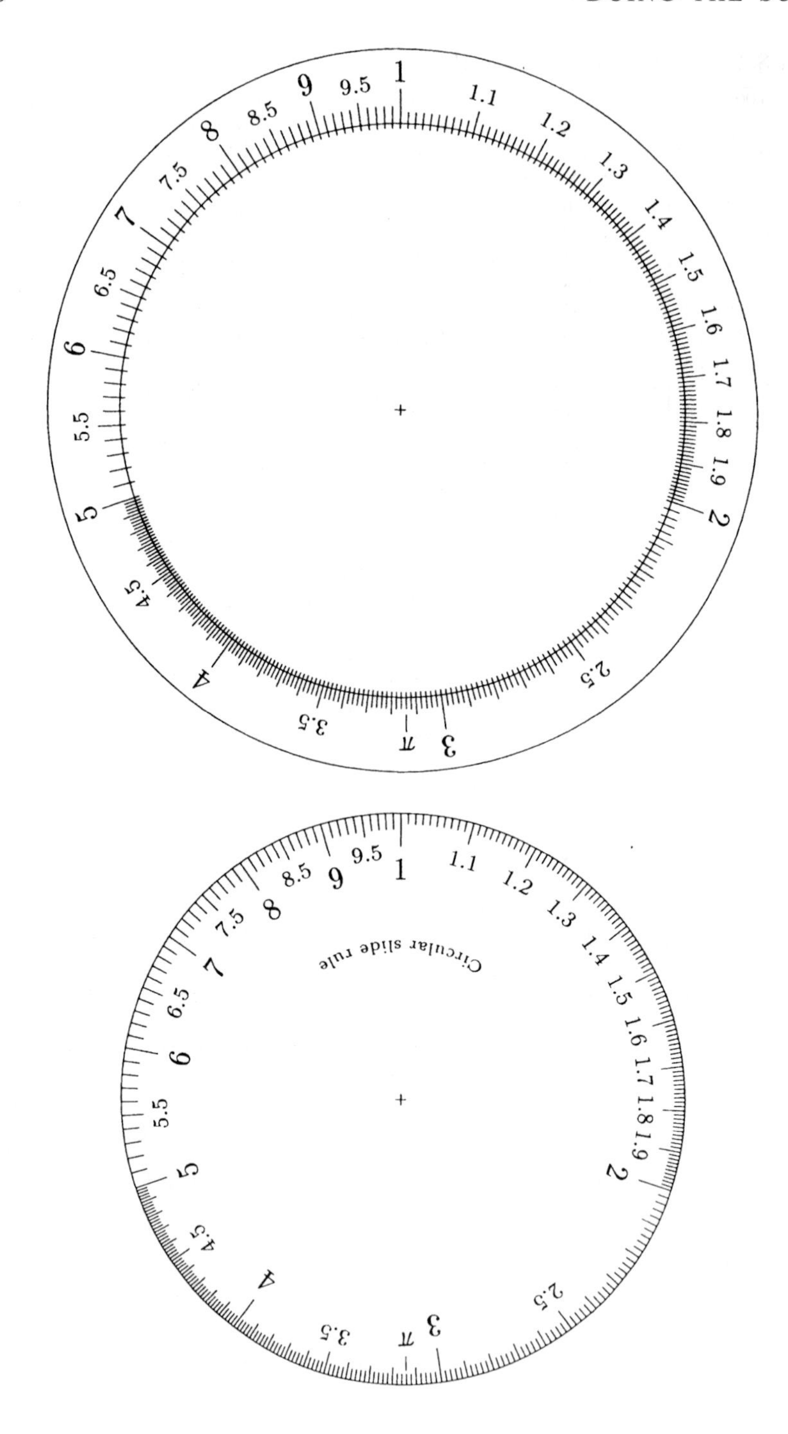
Circular slide rule
1 1.1 1.2 1.3 1.4 1.5 1.6 1.7 1.8 1.9 2 2.5 3 π 3.5 4 4.5 5 5.5 6 6.5 7 7.5 8 8.5 9 9.5

Table 8.2: Logarithms to the base 10.

	0.00	0.01	0.02	0.03	0.04	0.05	0.06	0.07	0.08	0.09
1.0	0.0000	0.0043	0.0086	0.0128	0.0170	0.0212	0.0253	0.0294	0.0334	0.0374
1.1	0.0414	0.0453	0.0492	0.0531	0.0569	0.0607	0.0645	0.0682	0.0719	0.0755
1.2	0.0792	0.0828	0.0864	0.0899	0.0934	0.0969	0.1004	0.1038	0.1072	0.1106
1.3	0.1139	0.1173	0.1206	0.1239	0.1271	0.1303	0.1335	0.1367	0.1399	0.1430
1.4	0.1461	0.1492	0.1523	0.1553	0.1584	0.1614	0.1644	0.1673	0.1703	0.1732
1.5	0.1761	0.1790	0.1818	0.1847	0.1875	0.1903	0.1931	0.1959	0.1987	0.2014
1.6	0.2041	0.2068	0.2095	0.2122	0.2148	0.2175	0.2201	0.2227	0.2253	0.2279
1.7	0.2304	0.2330	0.2355	0.2380	0.2405	0.2430	0.2455	0.2480	0.2504	0.2529
1.8	0.2553	0.2577	0.2601	0.2625	0.2648	0.2672	0.2695	0.2718	0.2742	0.2765
1.9	0.2788	0.2810	0.2833	0.2856	0.2878	0.2900	0.2923	0.2945	0.2967	0.2989
2.0	0.3010	0.3032	0.3054	0.3075	0.3096	0.3118	0.3139	0.3160	0.3181	0.3201
2.1	0.3222	0.3243	0.3263	0.3284	0.3304	0.3324	0.3345	0.3365	0.3385	0.3404
2.2	0.3424	0.3444	0.3464	0.3483	0.3502	0.3522	0.3541	0.3560	0.3579	0.3598
2.3	0.3617	0.3636	0.3655	0.3674	0.3692	0.3711	0.3729	0.3747	0.3766	0.3784
2.4	0.3802	0.3820	0.3838	0.3856	0.3874	0.3892	0.3909	0.3927	0.3945	0.3962
2.5	0.3979	0.3997	0.4014	0.4031	0.4048	0.4065	0.4082	0.4099	0.4116	0.4133
2.6	0.4150	0.4166	0.4183	0.4200	0.4216	0.4232	0.4249	0.4265	0.4281	0.4298
2.7	0.4314	0.4330	0.4346	0.4362	0.4378	0.4393	0.4409	0.4425	0.4440	0.4456
2.8	0.4472	0.4487	0.4502	0.4518	0.4533	0.4548	0.4564	0.4579	0.4594	0.4609
2.9	0.4624	0.4639	0.4654	0.4669	0.4683	0.4698	0.4713	0.4728	0.4742	0.4757
3.0	0.4771	0.4786	0.4800	0.4814	0.4829	0.4843	0.4857	0.4871	0.4886	0.4900
3.1	0.4914	0.4928	0.4942	0.4955	0.4969	0.4983	0.4997	0.5011	0.5024	0.5038
3.2	0.5051	0.5065	0.5079	0.5092	0.5105	0.5119	0.5132	0.5145	0.5159	0.5172
3.3	0.5185	0.5198	0.5211	0.5224	0.5237	0.5250	0.5263	0.5276	0.5289	0.5302
3.4	0.5315	0.5328	0.5340	0.5353	0.5366	0.5378	0.5391	0.5403	0.5416	0.5428
3.5	0.5441	0.5453	0.5465	0.5478	0.5490	0.5502	0.5514	0.5527	0.5539	0.5551
3.6	0.5563	0.5575	0.5587	0.5599	0.5611	0.5623	0.5635	0.5647	0.5658	0.5670
3.7	0.5682	0.5694	0.5705	0.5717	0.5729	0.5740	0.5752	0.5763	0.5775	0.5786
3.8	0.5798	0.5809	0.5821	0.5832	0.5843	0.5855	0.5866	0.5877	0.5888	0.5899
3.9	0.5911	0.5922	0.5933	0.5944	0.5955	0.5966	0.5977	0.5988	0.5999	0.6010
4.0	0.6021	0.6031	0.6042	0.6053	0.6064	0.6075	0.6085	0.6096	0.6107	0.6117
4.1	0.6128	0.6138	0.6149	0.6160	0.6170	0.6180	0.6191	0.6201	0.6212	0.6222
4.2	0.6232	0.6243	0.6253	0.6263	0.6274	0.6284	0.6294	0.6304	0.6314	0.6325
4.3	0.6335	0.6345	0.6355	0.6365	0.6375	0.6385	0.6395	0.6405	0.6415	0.6425
4.4	0.6435	0.6444	0.6454	0.6464	0.6474	0.6484	0.6493	0.6503	0.6513	0.6522
4.5	0.6532	0.6542	0.6551	0.6561	0.6571	0.6580	0.6590	0.6599	0.6609	0.6618
4.6	0.6628	0.6637	0.6646	0.6656	0.6665	0.6675	0.6684	0.6693	0.6702	0.6712
4.7	0.6721	0.6730	0.6739	0.6749	0.6758	0.6767	0.6776	0.6785	0.6794	0.6803
4.8	0.6812	0.6821	0.6830	0.6839	0.6848	0.6857	0.6866	0.6875	0.6884	0.6893
4.9	0.6902	0.6911	0.6920	0.6928	0.6937	0.6946	0.6955	0.6964	0.6972	0.6981
5.0	0.6990	0.6998	0.7007	0.7016	0.7024	0.7033	0.7042	0.7050	0.7059	0.7067
5.1	0.7076	0.7084	0.7093	0.7101	0.7110	0.7118	0.7126	0.7135	0.7143	0.7152
5.2	0.7160	0.7168	0.7177	0.7185	0.7193	0.7202	0.7210	0.7218	0.7226	0.7235
5.3	0.7243	0.7251	0.7259	0.7267	0.7275	0.7284	0.7292	0.7300	0.7308	0.7316
5.4	0.7324	0.7332	0.7340	0.7348	0.7356	0.7364	0.7372	0.7380	0.7388	0.7396
5.5	0.7404	0.7412	0.7419	0.7427	0.7435	0.7443	0.7451	0.7459	0.7466	0.7474
5.6	0.7482	0.7490	0.7497	0.7505	0.7513	0.7520	0.7528	0.7536	0.7543	0.7551
5.7	0.7559	0.7566	0.7574	0.7582	0.7589	0.7597	0.7604	0.7612	0.7619	0.7627
5.8	0.7634	0.7642	0.7649	0.7657	0.7664	0.7672	0.7679	0.7686	0.7694	0.7701
5.9	0.7709	0.7716	0.7723	0.7731	0.7738	0.7745	0.7752	0.7760	0.7767	0.7774

Table 8.3: Logarithms to the base 10.

	0.00	0.01	0.02	0.03	0.04	0.05	0.06	0.07	0.08	0.09
6.0	0.7782	0.7789	0.7796	0.7803	0.7810	0.7818	0.7825	0.7832	0.7839	0.7846
6.1	0.7853	0.7860	0.7868	0.7875	0.7882	0.7889	0.7896	0.7903	0.7910	0.7917
6.2	0.7924	0.7931	0.7938	0.7945	0.7952	0.7959	0.7966	0.7973	0.7980	0.7987
6.3	0.7993	0.8000	0.8007	0.8014	0.8021	0.8028	0.8035	0.8041	0.8048	0.8055
6.4	0.8062	0.8069	0.8075	0.8082	0.8089	0.8096	0.8102	0.8109	0.8116	0.8122
6.5	0.8129	0.8136	0.8142	0.8149	0.8156	0.8162	0.8169	0.8176	0.8182	0.8189
6.6	0.8195	0.8202	0.8209	0.8215	0.8222	0.8228	0.8235	0.8241	0.8248	0.8254
6.7	0.8261	0.8267	0.8274	0.8280	0.8287	0.8293	0.8299	0.8306	0.8312	0.8319
6.8	0.8325	0.8331	0.8338	0.8344	0.8351	0.8357	0.8363	0.8370	0.8376	0.8382
6.9	0.8388	0.8395	0.8401	0.8407	0.8414	0.8420	0.8426	0.8432	0.8439	0.8445
7.0	0.8451	0.8457	0.8463	0.8470	0.8476	0.8482	0.8488	0.8494	0.8500	0.8506
7.1	0.8513	0.8519	0.8525	0.8531	0.8537	0.8543	0.8549	0.8555	0.8561	0.8567
7.2	0.8573	0.8579	0.8585	0.8591	0.8597	0.8603	0.8609	0.8615	0.8621	0.8627
7.3	0.8633	0.8639	0.8645	0.8651	0.8657	0.8663	0.8669	0.8675	0.8681	0.8686
7.4	0.8692	0.8698	0.8704	0.8710	0.8716	0.8722	0.8727	0.8733	0.8739	0.8745
7.5	0.8751	0.8756	0.8762	0.8768	0.8774	0.8779	0.8785	0.8791	0.8797	0.8802
7.6	0.8808	0.8814	0.8820	0.8825	0.8831	0.8837	0.8842	0.8848	0.8854	0.8859
7.7	0.8865	0.8871	0.8876	0.8882	0.8887	0.8893	0.8899	0.8904	0.8910	0.8915
7.8	0.8921	0.8927	0.8932	0.8938	0.8943	0.8949	0.8954	0.8960	0.8965	0.8971
7.9	0.8976	0.8982	0.8987	0.8993	0.8998	0.9004	0.9009	0.9015	0.9020	0.9025
8.0	0.9031	0.9036	0.9042	0.9047	0.9053	0.9058	0.9063	0.9069	0.9074	0.9079
8.1	0.9085	0.9090	0.9096	0.9101	0.9106	0.9112	0.9117	0.9122	0.9128	0.9133
8.2	0.9138	0.9143	0.9149	0.9154	0.9159	0.9165	0.9170	0.9175	0.9180	0.9186
8.3	0.9191	0.9196	0.9201	0.9206	0.9212	0.9217	0.9222	0.9227	0.9232	0.9238
8.4	0.9243	0.9248	0.9253	0.9258	0.9263	0.9269	0.9274	0.9279	0.9284	0.9289
8.5	0.9294	0.9299	0.9304	0.9309	0.9315	0.9320	0.9325	0.9330	0.9335	0.9340
8.6	0.9345	0.9350	0.9355	0.9360	0.9365	0.9370	0.9375	0.9380	0.9385	0.9390
8.7	0.9395	0.9400	0.9405	0.9410	0.9415	0.9420	0.9425	0.9430	0.9435	0.9440
8.8	0.9445	0.9450	0.9455	0.9460	0.9465	0.9469	0.9474	0.9479	0.9484	0.9489
8.9	0.9494	0.9499	0.9504	0.9509	0.9513	0.9518	0.9523	0.9528	0.9533	0.9538
9.0	0.9542	0.9547	0.9552	0.9557	0.9562	0.9566	0.9571	0.9576	0.9581	0.9586
9.1	0.9590	0.9595	0.9600	0.9605	0.9609	0.9614	0.9619	0.9624	0.9628	0.9633
9.2	0.9638	0.9643	0.9647	0.9652	0.9657	0.9661	0.9666	0.9671	0.9675	0.9680
9.3	0.9685	0.9689	0.9694	0.9699	0.9703	0.9708	0.9713	0.9717	0.9722	0.9727
9.4	0.9731	0.9736	0.9741	0.9745	0.9750	0.9754	0.9759	0.9763	0.9768	0.9773
9.5	0.9777	0.9782	0.9786	0.9791	0.9795	0.9800	0.9805	0.9809	0.9814	0.9818
9.6	0.9823	0.9827	0.9832	0.9836	0.9841	0.9845	0.9850	0.9854	0.9859	0.9863
9.7	0.9868	0.9872	0.9877	0.9881	0.9886	0.9890	0.9894	0.9899	0.9903	0.9908
9.8	0.9912	0.9917	0.9921	0.9926	0.9930	0.9934	0.9939	0.9943	0.9948	0.9952
9.9	0.9956	0.9961	0.9965	0.9969	0.9974	0.9978	0.9983	0.9987	0.9991	0.9996
10	1.0000									

Index